Criti
and c
Proje

Fourth edition of
An Introduction to Critical Path Analysis

Critical Path Analysis and other Project Network Techniques

Keith Lockyer
BSc, CEng, FBIM, FIIM
Professor of Operations Management, University of Bradford

PITMAN PUBLISHING
128 Long Acre, London WC2E 9AN

Fourth edition 1984
Reprinted 1986, 1987
First published in Great Britain 1964
First published in paperback 1977

British Library Cataloguing in Publication Data
Lockyer, K.G.
Critical path analysis and other project network techniques.—4th ed.
1. Critical path analysis
I. Title II. Lockyer, K.G. Introduction to critical path analysis
658.4′032 TS158

ISBN 0–273–01951–1

Printed in Great Britain at The Bath Press, Avon

TO DORIS

Contents

The following are particular to:

Activity-on-arrow (AoA) systems: Chapters 2, 3, 6, 7, 8
Activity-on-node (AoN) systems: Chapters 4, 9, 10, 12

IMPORTANT NOTE

A *Solutions Manual* is available free of charge to lecturers, course tutors and practising managers on application to the publisher. Please write to the address below on official notepaper quoting teaching responsibilities or job title.

The *Solutions Manual* is not available to students.

Publisher
Professional & Business Studies
Pitman Books Ltd
128 Long Acre
London WC2E 9AN

Preface

The purpose of this text is, as with all previous editions, essentially a practical one. It is intended both for those who wish to assess the value of project network techniques (PNT) in dealing with their own problems, and also for those who, being convinced of its value, wish to put the technique into practice. It is therefore written 'from the ground up', assuming no prior knowledge on the part of the reader, and requiring no mathematical expertise. Experienced workers in this field may find that some of the discussion is apparently excessively long or excessively simple, but the subject matter has been shaped by the very large number of lectures and consultancy assignments undertaken by my colleagues and myself throughout the world. This work has highlighted the initial difficulties and problems encountered by students, and these have been elaborated upon as far as possible within a working book.

In a subject exploding as violently as PNT there is a temptation to include in a textbook every new idea, manipulation and treatment without regard to their practical value. Two tests have been applied to all new material: (1) Is it useful? (2) Is it proven? Failure to meet these criteria has caused material, however elegant or intellectually appealing, to be rejected. That material which serves only to provide good examination questions has been avoided.

This edition differs from previous editions in the following respects:

(1) *Activity-on-node (AoN) networking has been treated in its own right*. In previous editions a knowledge of AoN had to be derived from deductions from Activity-on-arrow (AoA) treatments. The increasing use of activity on node systems has made it desirable for a reader to be able to learn about AoN without reference to AoA.

(2) *Multiple dependency activity on node is discussed at some length*. It is suggested that this type of networking will become the most usual system in the near future. This is not a view shared by the author, but nevertheless there are sufficient workers in the field using this system to merit a substantial discussion.

(3) *Resource allocation is discussed comprehensively*. In practice, few fieldworkers will ever attempt to carry out a resource allocation

without recourse to an existing computer program. However, blind adherence to a program can lead to difficulties and it is felt that an understanding of some of the principles underlying resource allocation should be understood by serious workers. The methods suggested in the present book are not claimed necessarily to produce 'the best' answers. Indeed, the author believes that the best in this context is often unknown and unknowable. The imperfection of data and the variability of criteria are such that to hunt for an optimum is to hunt a chimera.

(4) *The use of cash as a resource is discussed more fully.* In practice it is not always easy to apply cost control to a network due to the organizational problems created by different control systems. Nevertheless, an increasing use is being made of cost control systems and it is believed that the discussion here will furnish an insight into the principles of cost control.

(5) *A set of questions is provided.* Present day economics prevent the publication of as comprehensive a set of questions as some readers would feel desirable. Nevertheless, PNT is essentially a subject where learning is best by doing, rather than reading, and it is hoped that the questions will meet some expressed need. A solutions manual can be obtained on demand from the publishers.

The subject matter being new and in a ferment, it is not surprising to find that existing practitioners are enthusiasts who are willing to discuss their experiences. I have derived much benefit from these discussions and should like to thank particularly David Armstrong (formerly of Richard Costain Ltd); Brian Fine (Fine & Curtis Ltd); Fred Moon (Southern Project Group, Central Electricity Generating Board); P. A. Rhodes (ICL Ltd); Jim Gordon (University of Birmingham) and Professor Geoffrey Trimble (University of Loughborough), from all of whom I have learned a great deal. It may well be that they will find some of their own ideas appearing in the following pages without appropriate acknowledgement, if so, it is quite unintentional and I should like to add my apologies to my thanks.

I should also like to thank those many readers who have written pointing out errors and obscurities. Any remaining faults, of course, are entirely my own responsibility.

K. G. Lockyer
September 1983

1 Introductory

When attempting to determine the completion date for any project, whether it be the building of a bridge, the mounting of a sales conference, the designing of a new piece of equipment or any other project, it is necessary to timetable all the activities that make up the task, that is to say, a *plan* must be prepared. The need for planning has always been present, but the complexity and competitiveness of modern undertakings now requires that this need should be met rather than just recognized. The first attempt at a formal planning system was the *Gantt chart*.

The Gantt chart

In this type of chart, the time that an activity should take is represented by a horizontal line, the length of the line being proportional to the duration time of the activity. In order that several activities can be

Time / Activity	Week number																	
	1	2	3	4	5	6	7	8	9	10	11	12	13	14	15	16	17	18

Fig. 1.1

represented on the same chart, a framework or ruling is set up, giving time flowing from left to right, the activities being listed from top to bottom (see Fig. 1.1).

Assume for the sake of simplicity that there are three activities, *A*, *B* and *C*, which must be carried out in sequence, and that the duration times are:

Activity *A*: 4 weeks
Activity *B*: 6 weeks
Activity *C*: 5 weeks

This is represented on the Gantt chart as shown in Fig. 1.2 and reveals quite clearly how work should progress. Thus, by the end of week 8, the whole of activity *A* and two-thirds of activity *B* should be complete.

Time / Activity	Week number																	
	1	2	3	4	5	6	7	8	9	10	11	12	13	14	15	16	17	18
A																		
B																		
C																		

Fig. 1.2

To show how work is actually progressing, a bar or line can be drawn within the uprights of the activity symbol, the length of the bar representing the amount of work completed. Thus, if 50 per cent of an activity is complete, then a bar half the length of the activity symbol is drawn (Fig. 1.3). This gives a very simple and striking representation of work done, particularly if a number of activities are represented on the same chart, as shown previously.

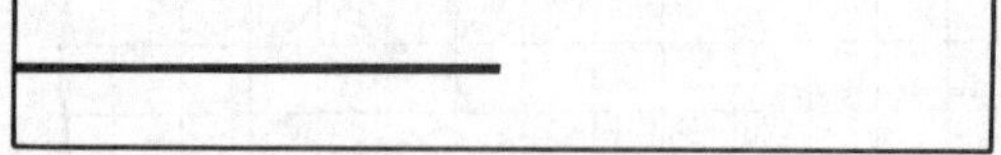

Fig. 1.3

If this chart has been correctly filled in, and it is viewed at the end of week 7 (denoted by two small arrows at the top and the bottom of the chart as shown in Fig. 1.4), then the following information is readily apparent:

Activity *A* should be complete and, in fact, is so.
Activity *B* should be 50 per cent complete, but, in fact, is only 17 per cent finished.

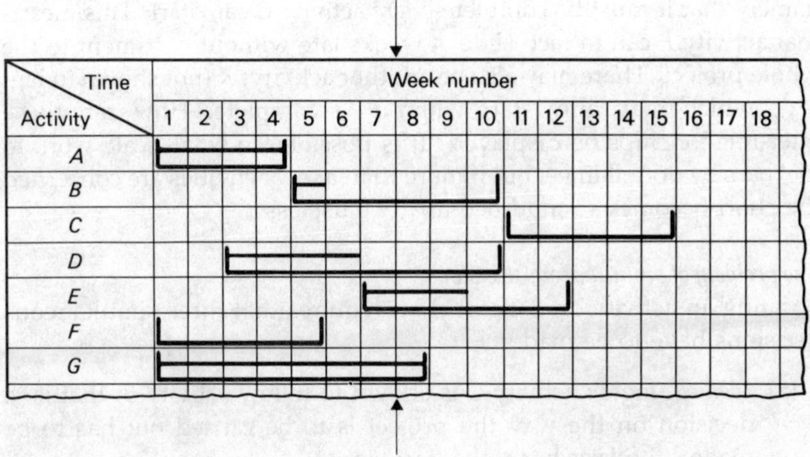

Fig. 1.4

Activity *C* should not be started and, in fact, is not started.
Activity *D* should be 62 per cent complete and, in fact, is only 50 per cent finished.
Activity *E* should be 17 per cent complete and, in fact, is 50 per cent finished.
Activity *F* should be complete and, in fact, is not started.
Activity *G* should be 87 per cent complete and, in fact, is complete.

Thus we see that incomplete bars to the *left* of the cursor mean under-fulfilment, while those to the *right* mean over-fulfilment. By the use of codes and/or symbols, the reasons for any delays can be displayed, and the whole chart can be very succinctly informative, combining both planning and recording progress. For many tasks the Gantt chart is unsurpassed, and its use has been very highly developed.

The difficulties with the Gantt chart

Though valuable, the Gantt chart presents two major difficulties, one concerning the problem of inter-relationships, the second that of needing to take several decisions simultaneously.

The problem of inter-relationships
Consider activity *F* in Fig. 1.4. It is shown here to start at the beginning of the project. However, it may be that there is another requirement,

namely that it must be complete *before* activity *C* can start. This means that activity *F* can in fact 'slide' 5 weeks late without detriment to the whole project. There may also be another activity, *K* (not shown in Fig. 1.4), which may only start when *F* is complete. How can these inter-relationships be displayed? It is possible, in small-scale work to 'tie' bars by dotted lines, but if more than a few activities are concerned the chart becomes so muddled as to be useless.

The problem of simultaneous decisions

Locating an activity on a Gantt chart requires that three simultaneous decisions have to be made:

(1) *Method* (logic): activity *C* is shown to follow activity *B*, that is, a decision on the way the project is to be carried out has to be made—the *logic* has to be decided.
(2) *Time:* any activity bar has a *length*, that is, a decision has been made on the time that each activity will occupy.
(3) *Resources:* locating an activity in a position implies that resources are available to carry out the activity.

All projects have these three dimensions—*logic, time, resources*—and each is equally important. To require a planner to make decisions on these three features at one time is to set an impossible task, yet this is what is required when a Gantt chart is drawn. The decisions must be taken serially.

The characteristics of effective planning

Present decisions affect both present *and future* actions, and if immediate, short-term decisions are not taken within the framework of long-term plans, then the short-term decisions may effectively impose some long-term actions that are undesirable but inescapable. Military writers categorize these long- and short-term plans as:

> *Strategical* plans, which are those made to 'serve the needs of generalship'.
> *Tactical* plans, which are those made 'when in contact with the enemy'.

Ideally, of course, strategical plans are made before the start of an operation and, by following them, the operation is successfully concluded. Inevitably, however, tactical decisions will have to be made and these can only be successful if they are made within the context of the strategic plan.

To permit effective tactical decision-making, it is necessary for the strategy to be expressed in a form that is:

(1) Explicit.
(2) Intelligible.
(3) Capable of accepting change.
(4) Capable of being monitored.

It is suggested that these are the minimum acceptable characteristics of an effective plan. Given these characteristics, tactical decisions can then be taken which have results that are both *predictable* and *acceptable* to the strategic needs.

It must be realized that there is no absolute definition of strategy and tactics: at any one level in a hierarchy a tactical plan should be made to fit the needs of the strategy received from a higher level, and this tactical plan then becomes the strategy for a lower level. Freedom to make appropriate decisions must be given to those who will be held responsible for performance, and if these decisions are to be meaningful in a larger context, then it is imperative that this larger context shall be known. Freedom 'within the law' is as important a concept in management as in the community. Too often, tactical decisions are taken to meet the needs only of immediate expediency, and the results may well be disastrous in the long term.

An effective plan will be:

(1) Explicit.
(2) Intelligible.
(3) Capable of accepting change.
(4) Capable of being monitored.

Project network techniques (PNT) as a planning tool

The middle 1950s and the 1960s saw an explosion of interest in the problems of planning, and the family of methods, *project network techniques* (PNT), was born and very rapidly developed. Essentially, these techniques involve representing the proposed project by a diagram (or 'model') built up from a series of *arrows* and *nodes* (boxes or circles). The original structure of the model depends only upon the proposed method of proceeding and it is drawn in such a way that the logic is easily displayed and tested (a network planner will often be heard to mutter '*That* must follow *that*, and *that* must follow *that* . . . but why have I shown *that* starting before *that* . . . where's my rubber—I'll alter that!').

Once the arrow diagram is believed to show an acceptable logic, times are set to the various constituents of the diagram. A calculation is then carried out to discover the total time for the project. If this is satisfactory (oh glorious day!) then no further action is necessary at this stage. If the total project time needs to be reduced then the activities dictating this time—'the critical activities'—are examined to discover whether they can be adequately shortened by using a different method, or by changing the logic of the network itself. This exchange between logic and time continues until an acceptable situation is obtained.

Time and logic having been considered, it may be necessary—in some fortunate cases it is not necessary—to consider the resources required by the plan as it now stands and the resources available. This is done by moving through the network and adding up ('aggregating') the resources for each period of time. The aggregated resources ('loads') are compared with the resources available ('capacity'). If they exceed the resource ceiling then the network is re-examined to see if any manipulation can take place to 'spread' the load satisfactorily. If not, then clearly either the available resources or the total project time (*TPT*) or both must be increased.

One of the very useful by-products of drawing a network is that the diagram can be used as a means of communication. It can, for example, record decisions on how a project is to be completed, or it can enable an executive to pass on information to his successor or his subordinates. It is quite certain that the network will be seen and used by persons other than those who prepared it. With this 'communicating' aspect in mind, it is important from a practical point of view that activities should be *unequivocal* statements in *positive* terms that have *significance* within the context of the task being considered. Furthermore, wherever possible, constituents should be chosen in such a way that the responsibility for carrying out the activity can be explicitly assigned.

Two types of PNT

Broadly, there are two PNT families, the activity-on-arrow (AoA) family, where an activity is represented by arrow, and the activity-on-node (AoN) family, where an activity is represented by a box or node. These two families each have their own advantages and their own adherents. The author has used both extensively and believes that neither shows an overwhelming advantage over the other. Both have the characteristics of a good plan—namely, they are:

Explicit	— often uncomfortably so: their clarity is uncompromising.
Intelligible	— with education they are very intelligible—it is often easy to deduce the way in which the planner's mind was working.
Capable of accepting change	— changes are easily made to a network: a working network that does not incorporate change is a rarity.
Capable of being monitored	— the progress of the project as a whole and of the constituent parts can be very easily deduced *providing that the information on performance is fed back to the network appropriately fast*.

Where PNT can be used

PNT can be used in situations where the start and the finish of the task can be identified: continuous or flow production is not susceptible to planning by PNT *although the setting-up or pre-production work is*. The size of the project is of no consequence—networking has been used to plan a simple test procedure just as successfully as to plan the construction of a town or the launching of a space ship.

It is quite impossible to list all the applications of networks since the technique is now used extremely widely. However, to give some idea of 'spread', the following is a brief summary of some uses where PNT has been used to plan and control the use of time, materials and resources of which the author or his colleagues have personal knowledge. It must be emphasized that this list is not exhaustive: new applications are continually being found.

(1) *Overhaul:* plant, equipment, vehicles and buildings, both on a routine and an emergency basis.
(2) *Construction:* houses, flats and offices, including all pre-contract, tendering and design work.
(3) *Civil engineering:* motorways, bridges, road programmes, including all pre-contract, tendering and design work.
(4) *Town planning:* control of tendering and design procedures and subsequent building and installation of services.
(5) *Marketing:* market research, product launching and the setting-up and running of advertising campaigns.
(6) *Ship building:* design and production of ships.
(7) *Design:* design of cars, machine tools, guided weapons, computers, electronic equipment.
(8) *Pre-production:* control of production of jigs, tools, fixtures and test equipment.
(9) *Product change-over:* the changing over from one product or family of products to another, for example, 'winter' to 'summer' goods.
(10) *Commissioning and/or installation:* power generation equipment of all types, and data processing plant.
(11) *Modification programmes:* the modification of existing plant or equipment.
(12) *Office procedures:* investigations into existing administrative practices (for example, the preparation of monthly accounts) and the devising and installing of new systems.
(13) *Consultancy:* the setting up and control of consultancy assignments.

Introducing PNT into an organization

As with any other new managerial tool, PNT will require to be introduced into an organization with care. It is suggested that the following points should be observed:

(1) PNT is not a universal tool—there are situations where it cannot be usefully employed. These situations are, in general, those where activity is continuous, for example, flow production. A PNT-type situation is characteristically one that has a definable start and a definable finish.
(2) PNT is not a panacea—it does not cure all ills. Indeed, *in itself* it does not solve any problems, but it does expose situations in a way that will permit effective examinations both of the problems

and of the effect of possible solutions. However, the formulating and implementing of any solution will remain the responsibility of the appropriate manager.

(3) Networking must not be made a mystery, known only to a chosen few. All levels require to appreciate the method and its limitations, and an extensive educational programme will be necessary to ensure that knowledge is spread as widely as possible.

(4) The person initiating networking into an organization must be of sufficient stature and maturity to be able to influence both senior and junior personnel.

(5) Wherever possible, the early applications of PNT should be to simple situations. If networking is first employed on a very difficult task it may fail, not because of the difficulty of networking but because of the difficulty of the task itself. However, the failure is likely to be attributed to PNT and the technique will be discredited.

(6) The first application of networking will undoubtedly excite considerable interest, and attract many resources, both physical and managerial. This may starve other non-PNT-planned tasks to their detriment, and it may produce exceptionally good results on the 'PNT' job that cannot be reproduced on later jobs. While it is impossible to avoid this 'halation' completely, its existence needs to be recognized.

(7) PNT will involve committal to, and the acceptance of, responsibilities expressed in quantitative terms. Many supervisors find this difficult to accept, and will often try to escape by creating unreal problems. It is vital to make it quite clear that PNT is not a punitive device: it is a tool to assist, not a weapon to assault.

(8) The purpose for which the network is to be used must be known in advance. A 'feasibility' network is likely to be different to a working network. A network designed to control at one level in the organization will differ from that used to control at a different level.

(9) Excessive detail must be eschewed. The author has found, both in his own work and in talking to hundreds of network planners that *excessive detail in the network is the most common cause of failure.* A network with more than a very few hundred activities—say 400 at most—is likely to be excessively detailed. Such networks should be broken down into smaller sub-networks, possibly building a hierarchy of networks.

2 The elements of an activity-on-arrow network: I General

As already stated, a project is represented by an *arrow diagram*, which is not unlike the work study engineer's flow chart. In the activity-on-arrow (AoA) system, more generally known as critical path analysis (CPA) or PERT, the arrow diagram is made up of only two basic elements:

(1) *An activity*, which is an element of the work entailed in the project. In some instances the 'work' is not real in the sense that neither energy nor money is consumed, and in some cases (see dummy activities below) no time is used. However, ignoring these last cases, an activity is a task that must be carried out. Thus, 'waiting delivery of component *X*' is an activity just as much as is 'making component *Y*', since both are tasks that must be carried out. This 'non-work' aspect of some activities is sometimes found difficult to accept until the test of *needfulness to the project* is applied. Once this test is applied it is clear that waiting for delivery is an activity in the sense in which the word is used in drawing networks.

(2) *An event*, which is the start and/or finish of an activity or group of activities. The essential criterion is that a definite, unambiguous point in time can be isolated—a broad band of availability is of no use. The word 'event' may be misleading here, since there may in fact be a concurrence of a number of separate events, and for this reason some authorities prefer the terms 'node', 'junction', 'milestone' or 'stage'. In general, 'milestone' is reserved for particularly significant events that require special monitoring. 'Node' is possibly the most generally used term, and will be used subsequently. It must be remembered that a 'node' in AoA has a different significance to a 'node' in activity-on-node (AoN).

Conventions adopted in drawing networks in AoA

There are only two conventions usually adopted in drawing networks and, like all conventions, they may be ignored if circumstances warrant. In the early stages of network drawing, it is suggested that the conventions be respected until sufficient experience has been gained to justify dropping them. The conventions are:

(1) *Time flows from left to right.*

(2) *Head nodes always have a number higher than that of the tail node.* This allows activities to be referred to simply and succinctly by their tail and head numbers, so that 'activity 3–4' means only 'the activity that starts from node 3 and proceeds to node 4'; it *cannot* mean 'the activity that starts from node 4 and finishes at node 3'. Some computer programs exist that do not require this convention to be followed, but experience will show that it is nevertheless a useful one, as discussed in the section on 'looping' on page 16. It will also be found to 'sound' better.

It may be convenient to remark here that it is not necessary for all numbers to be in sequence, that is, that numbers need not follow each other in natural order. In fact it is sometimes useful, when numbering events, to leave gaps in the normal sequence so that, if it is necessary to modify a drawing, it is not also necessary to renumber all nodes—a tedious task. This may then result in nodes being numbers 1, 2, 3, 7, 14, 15, 18, 19, 20 and so on, the numbers 4, 5, 6, 8, 9, 10, 11, 12, 13, 16, 17 not appearing at all. No inconvenience will be found to result from this, and it is the author's practice initially to number the network using only multiples of five. It is useful to realize that the head and tail numbers of the activities effectively specify the logic of the diagram, and that from a list of head and tail numbers, the network can be constructed.

The graphical representation of events and activities in AoA

Events are represented by numbers, usually within convenient geometrical shapes—often circles. Activities are represented by arrows, the arrow-heads being at the completion of the activities. The length and orientation of the arrow are of no significance whatsoever, being chosen only for convenience of drawing. The activity of digging a hole can equally well be represented by Figs 2.1–2.5.

All of these have, within an arrow diagram, precisely the same significance, namely that to proceed from event 1 to event 2 it is necessary to carry out activity 1–2. It is equally not essential that arrows should be straight, although it will be found that the appearance of the whole diagram will be improved if the main portion of each arrow is

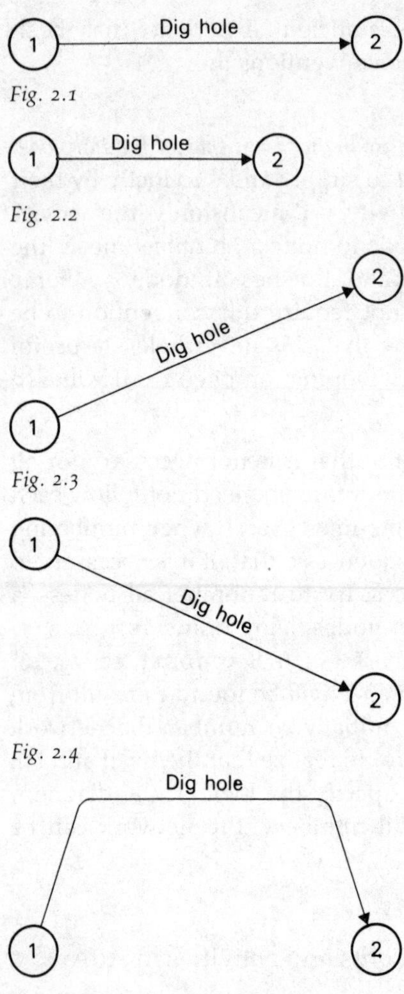

Fig. 2.1

Fig. 2.2

Fig. 2.3

Fig. 2.4

Fig. 2.5

both straight and parallel to the main horizontal axis of the paper on which the diagram is drawn. This will often require that arrows are 'bent', as in the last sketch above. The description of the activity should always be written upon the straight portion of the arrow.

It is strongly recommended that wherever possible this method of drawing should be adopted. There is a temptation to substitute a code letter for an activity description when drawing a network. This should be resisted at all costs as it makes the checking of the network extremely difficult if not impossible. It also destroys the 'communicating' ability of the network.

The representation of time in AoA networks

The expected time that will be required to complete an activity (the 'duration time') is written as a central subscript to the activity. Thus, if it is anticipated that it will require 6 days to dig a hole the activity would be as shown in Fig. 2.6. It should be noted that *more* than 6 days might be *available* for digging. This matter will be dealt with later in Chapter 7.

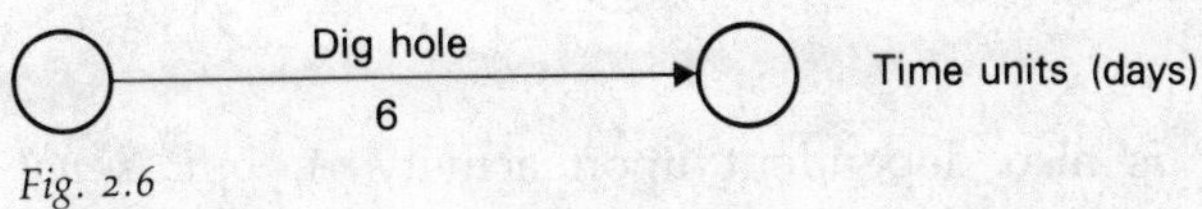

Fig. 2.6

If an activity must, for some reason external to the network, be completed by a given date, then an inverted triangle can be drawn at the head of the activity arrow. For example, if the digging of the hole must be completed by day 20, then the diagram would be as shown in Fig. 2.7.

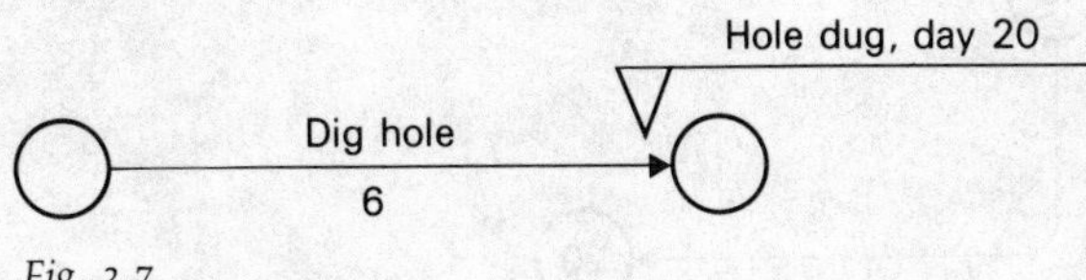

Fig. 2.7

Identification of activities

The node at the beginning of an activity is known as a 'tail' or 'preceding' node, while that at the conclusion of an activity is known as a 'head' or 'succeeding' node. Some writers refer to tail and head nodes as *i* and *j* nodes, this deriving from the generalization of an activity as in Fig. 2.8.

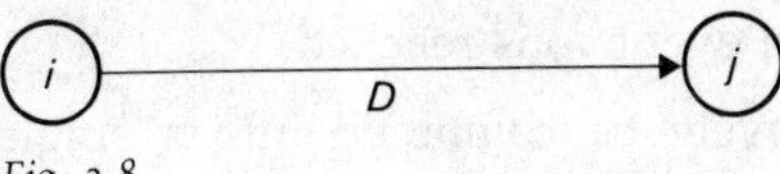

Fig. 2.8

This usage is extremely convenient when drawing up tables, where the single letters *i* and *j* are simpler to use than the words 'preceding' and 'succeeding', as recommended in BS 4335: 1972, or 'tail' and 'head'.

Fundamental properties of events and activities

Basically, the representation of events and activities is governed by one, simple, *dependency rule* which requires that an activity which depends upon another activity is shown to emerge from the head of the activity upon which it depends, and that only dependent activities are drawn in this way. Thus, if activity *B* depends upon activity *A*, then the two activities are drawn (Fig. 2.9)

Fig. 2.9

while if activity *C* is also dependent upon activity *A*, but is *not* dependent upon activity *B*, then the three activities are drawn (Fig. 2.10).

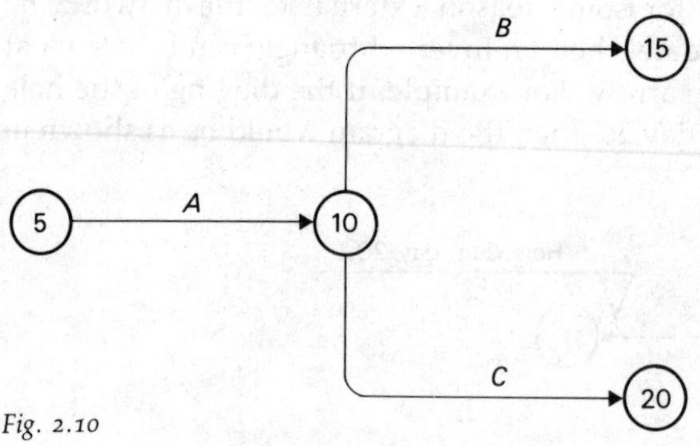

Fig. 2.10

This dependency rule gives rise to two fundamental properties of events and activities:

(1) *An event cannot be said to occur* (or 'be reached' or 'be achieved') *until all activities leading into it are complete.* For example, if a network such as Fig. 2.11 is found then event 10 can only be said to occur when activities 3–10, 4–10 and 5–10 are all complete.

(2) *No activity can start until its tail event is reached.* Thus, in Fig. 2.12 activity 10–11 cannot start until event 10 is reached.

These two statements can effectively be combined into a single comment, namely that 'No activity may start until all previous activities in the same chain are complete'. It must be understood, however, that this single statement has two facets as set out in (1) and (2) above. A crude but useful informality is: 'tail depends on head, tail does not depend on tail'.

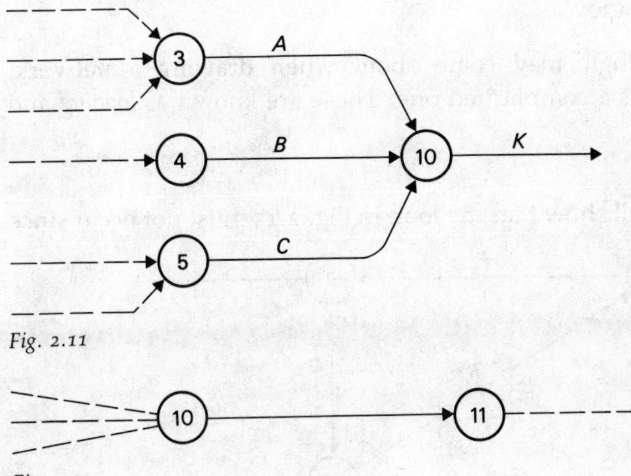

Fig. 2.11

Fig. 2.12

'Merge' and 'burst' nodes

Events into which a number of activities enter and one (or several) leave (Fig. 2.13) are known as 'merge' nodes. Events that have one (or several) entering activities generating a number of emerging activities are known as 'burst' nodes (Fig. 2.14).

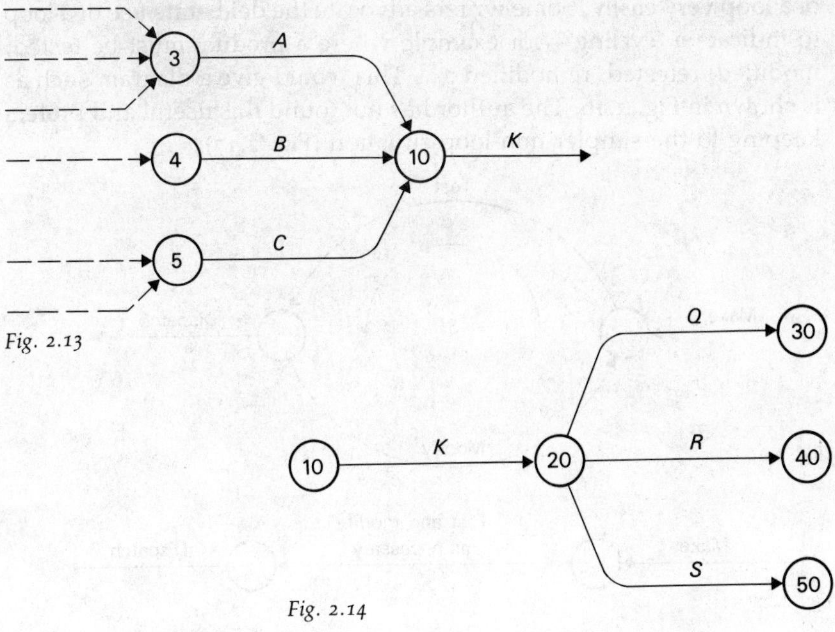

Fig. 2.13

Fig. 2.14

Two errors in logic

Two errors in logic may come about when drawing a network, particularly if it is a complicated one. These are known as *looping* and *dangling*.

Looping

Consideration will show that the loop in Fig. 2.15 must not occur since

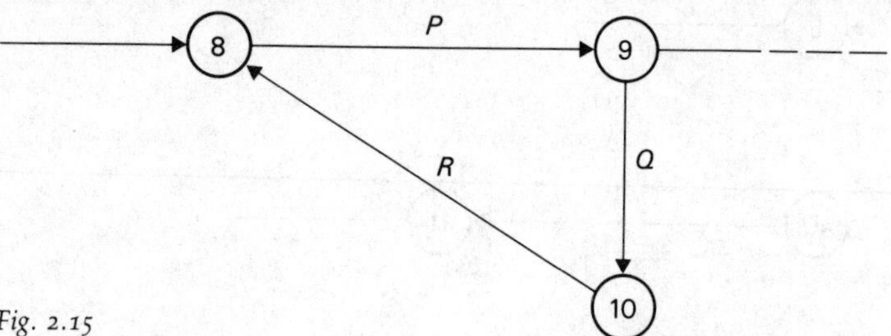

Fig. 2.15

this would represent an impossible situation: 'activity *R* depends on activity *Q* which depends on activity *P* which depends on activity *R* which depends on activity *Q* . . .'. If looping like this appears to arise, the logic underlying the diagram must be at fault, and the construction of the diagram must be re-examined. Adherence to the convention that no activity can start until its tail event is reached reveals the existence of a loop very easily. Some writers advocate the deliberate use of a loop to indicate a 'cycling'—for example where a product must be tested, modified, retested, remodified . . . This would give a diagram such as is shown in Fig. 2.16. The author has not found this useful and prefers keeping to the simpler non-loop situation (Fig. 2.17).

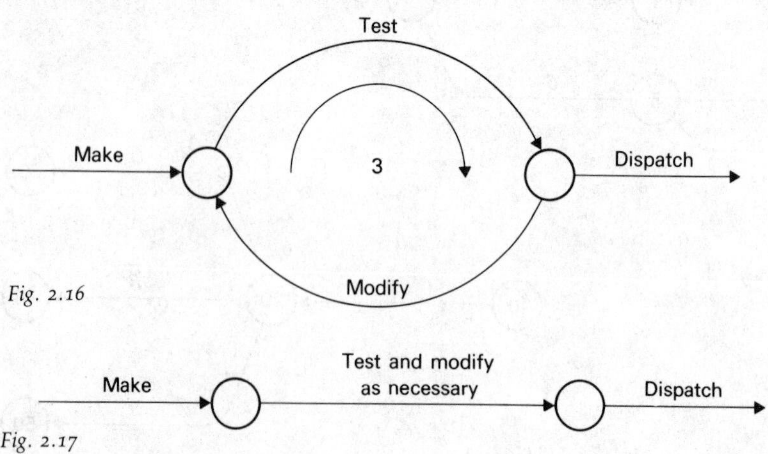

Fig. 2.16

Fig. 2.17

Dangling

Similarly, the situation represented by Fig. 2.18

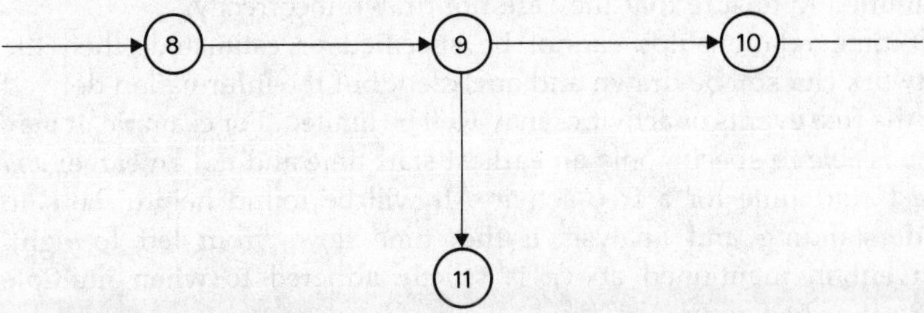

Fig. 2.18

is equally at fault, since the activity represented by the dangling arrow 9–11 is undertaken with no result. Such arrows often result from hastily inserted afterthoughts. Two rules can be enunciated which, if followed, will avoid dangling arrows, namely, 'all nodes, except the first and the last, must have at least one activity entering and one activity leaving them' and 'all activities must start and finish with a node'. There are special occasions when 'dangling' activities can be accepted but any appearance of a 'dangle' should be very carefully considered to ensure that it does not arise from an error in logic or an inadequate understanding of the task being considered. See 'multiple starts and finishes' below.

When set out in isolated form as above, both errors are quite obvious. However, in a complex network these errors (particularly looping) can arise, a loop for example forming over a very large chain of activities. Before finalizing on a network it is wise to examine for both the above logical errors. If the network is being processed by computer, the computer program itself will probably have preliminary tests written into it to check for looping and dangling before any further calculations are undertaken, and appropriate error signals will be generated.

Multiple starts and finishes

There are occasions when a network can have more than one start and/or finish. These usually arise when some activities are entirely independent of the control of the user of the network. Wherever possible, these multiple starts/finishes should be 'tied' into the network by activity arrows giving the desirable estimated or necessary time relationships between the 'free' event and the rest of the network. The existence of these free events should, of course, be very carefully examined to ensure that they are not drawn incorrectly.

If time relationships cannot be specified or estimated, then the network can still be drawn and analyzed, but the information derived for the free events or activities may well be limited. For example, it may be possible to specify only an earliest start time and not an earlier *and* latest start time for a free activity. It will be found helpful both to understanding and analysis if the 'time flows from left to right' convention mentioned above is strictly adhered to when multiple starts/finishes occur.

3 The elements of an activity-on-arrow network: II Dummies

Dummy activities

In some cases it is necessary to draw 'dummy' activities, that is, activities that do not require either resources or time. These are usually drawn as broken arrows:

although some workers use a solid arrow with a subscript *D*, for dummy:

D

or a subscript *O*, indicating the need for zero resources and zero time:

O

but the last two are to be discouraged. However drawn, a dummy activity *is always subject to the basic dependency rule* that an activity emerging from the head of another activity depends upon that activity.

There are two occasions when dummies are used:

(1) Identity dummies.
(2) Logic dummies.

Identity dummies

When two or more parallel independent activities have the same head and tail events, the identity of the activities, as given by the event numbers, could be lost. For example, if in making a cup of instant coffee two activities 'boil water' and 'heat milk' could proceed simultaneously, then the diagram as shown in Fig. 3.1 might appear. This would result in two activities having the same head and tail number.

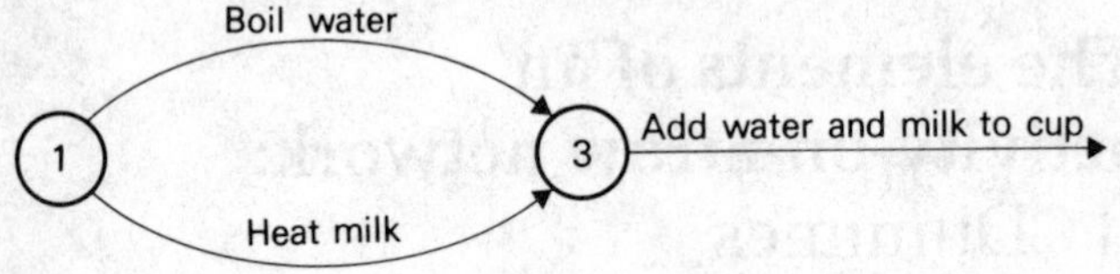

Fig. 3.1

To avoid the resultant confusion a dummy activity is introduced, which can be either activity 1–2 (Fig. 3.2) or activity 2–3 (Fig. 3.3).

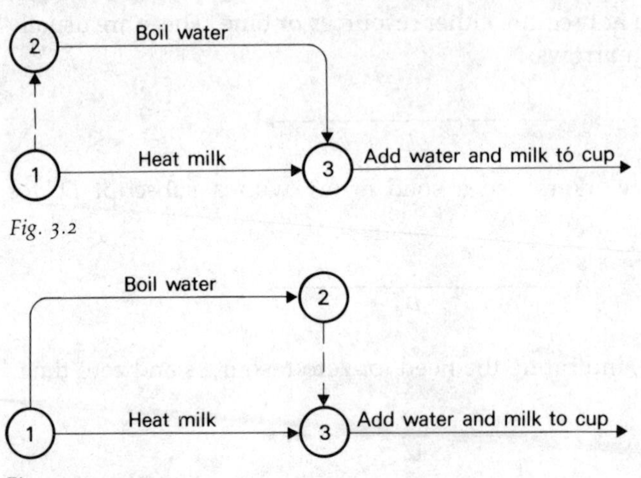

Fig. 3.2

Fig. 3.3

Note: In Fig. 3.3 the 'boil water activity' has been opened to accept the dummy. The other activity could equally well have been chosen, giving four possible diagrams. Which activity is in fact broken is a matter of indifference, though it is useful to insert the dummy at the tail (Fig. 3.2) rather than at the head of an activity.

Logic dummies

When two chains have a common event yet they are in themselves wholly or partly independent of each other, then an error in logic could unwittingly arise. Consider the situation:

Activity *K* depends on activity *A*. (1)
Activity *L* depends on activities *A* and *B*. (2)

At first sight the diagram might appear to be as shown in Fig. 3.4. Unfortunately an error is displayed by this diagram. Activity *L* is, quite correctly, shown to be dependent on activities *A* and *B*. However, activity *K* is also shown to be dependent on both activities, whereas it depends only on activity *A*. To resolve this a new activity (a dummy) is introduced to separate *K* from *B* (Fig. 3.5).

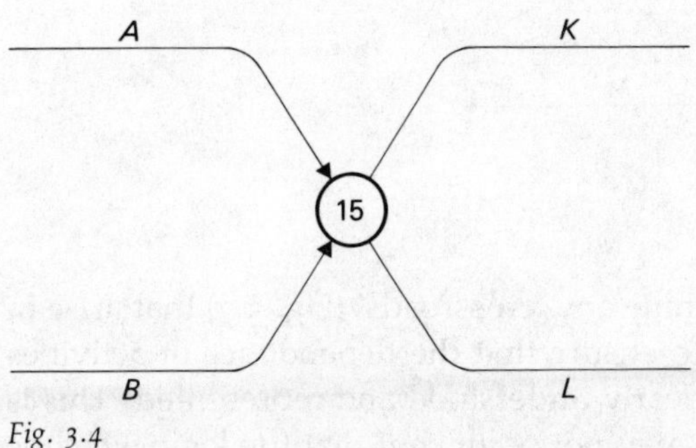

Fig. 3.4

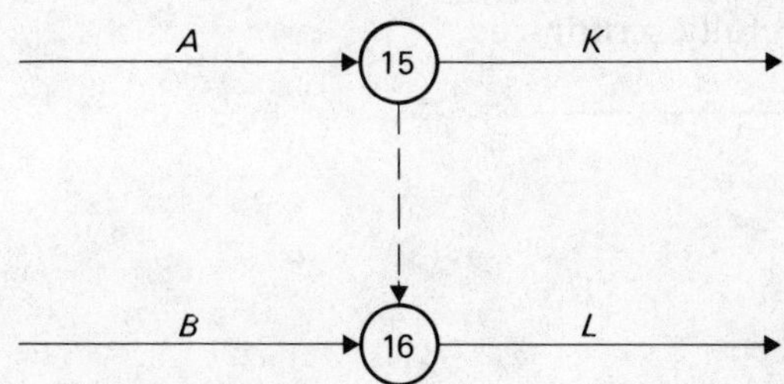

Fig. 3.5

Examining Fig. 3.5 it will be seen that:

K depends on *A*.
K does not depend on the dummy.

Hence *K* depends only on *A*. (1)

L depends on *B*.
L does depend on the dummy.
The dummy depends on *A*.

Hence *L* depends on *A* and *B*. (2)

(1) and (2) are the situations which it is required to represent.

It must be noted that multiple dummies may be necessary to maintain logic. For example, the situation:

Activity *K* depends on activity *A*
Activity *L* depends on activities *A* and *B*
Activity *M* depends on activity *B*

is represented by the diagram given in Fig. 3.6.

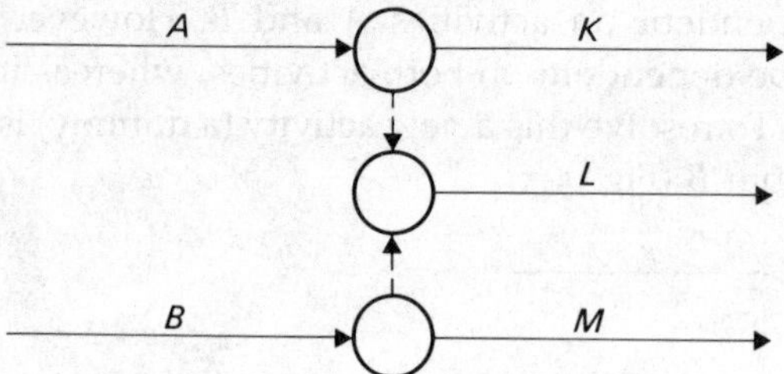

Fig. 3.6

Helpful hint

It is highly desirable to examine any 'crossroads' (Fig. 3.7) that arise in the drawing of a network to ensure that the dependence of activities upon one another is quite clearly understood and represented. This is not to say that 'crossroads' may not occur, but that the logic which is displayed must be very carefully scrutinized.

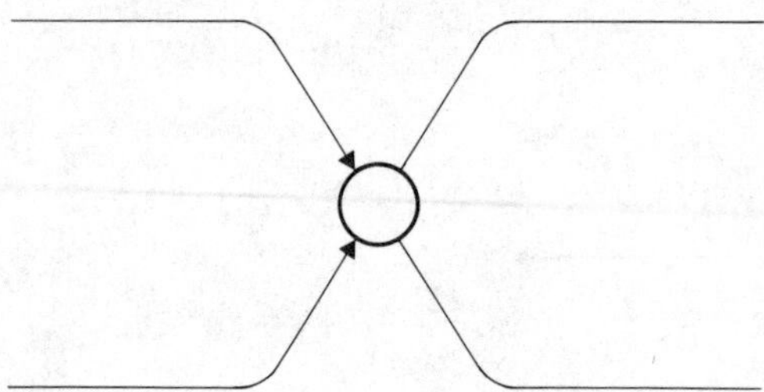

Fig. 3.7

The direction of dummies

Trouble is often encountered in assigning a direction to a dummy activity. If the purpose of the dummy is quite clearly understood, then the direction of the dummy becomes clearer. Thus if, in Fig. 3.5, the dummy exists to release activity *K* from activity *B* then the dummy emerges from the tail node of activity *K*; on the other hand, if activity *K* depended on activities *A* and *B* and activity *L* depended only on activity *B* and *not* on activities *A* and *B*, the general configuration would remain unaltered except that the dummy arrow-head would point the other way, i.e. from event 16 to event 15. Reference should be made to an earlier discussion on 'fundamental properties of events and activities' (page 14), and it should be clearly understood that the situation shown in Fig. 3.8 represents activity *X* being dependent on both activity *R* and activity *S*, while Fig. 3.9 represents activity *X* being dependent only on activity *R* and *independent* of activity *S*.

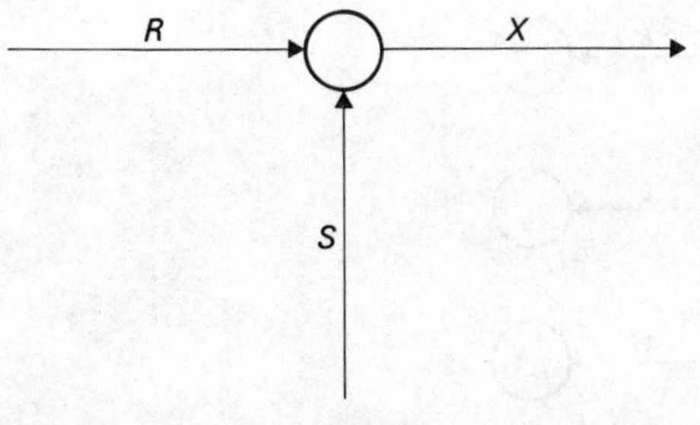

Fig. 3.8

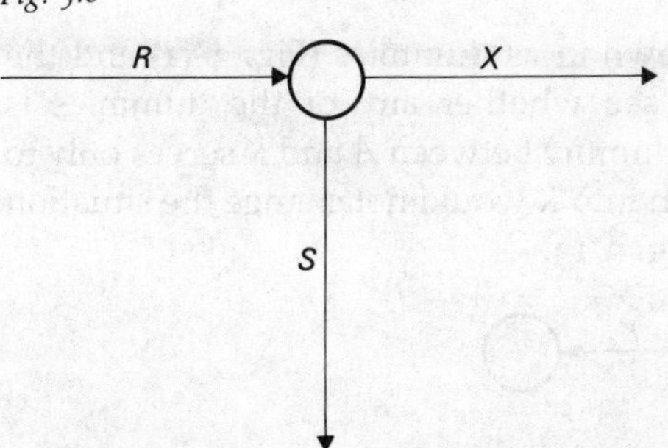

Fig. 3.9

The author has found it useful to consider the dummy as a 'one-way dependency street'—dependency can 'flow' from the tail of the dummy to its head, it cannot flow from its head to its tail.

The location of dummies

Difficulty is sometimes experienced in correctly locating dummies, a difficulty that diminishes with experience. A simple, albeit sometimes tedious, method of locating dummies, first introduced to the author by Norman Raby, is to draw *all* dependencies as dummies, and then to remove any redundant dummies. Consider the situation (assumed to be in the middle of a network) of the five dependencies:

Activity *K* depends on activity *A*.
Activity *L* depends on activities *A* and *B*.
Activity *M* depends on activities *B* and *C*.

The six activities, complete with head and tail events, are drawn. If they can be roughly located in the correct position, it will be found helpful but not essential (Fig. 3.10).

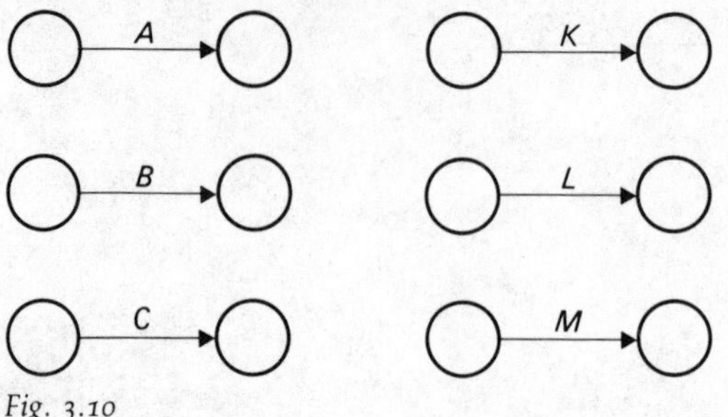

Fig. 3.10

All five dependencies are drawn in as dummies (Fig. 3.11) and the diagram is then examined to see whether any of the dummies is unnecessary. For example, the dummy between *A* and *K* serves only to 'extend' *K*, and its amalgamation into *K* would not change the situation so that Fig. 3.12 can become Fig. 3.13.

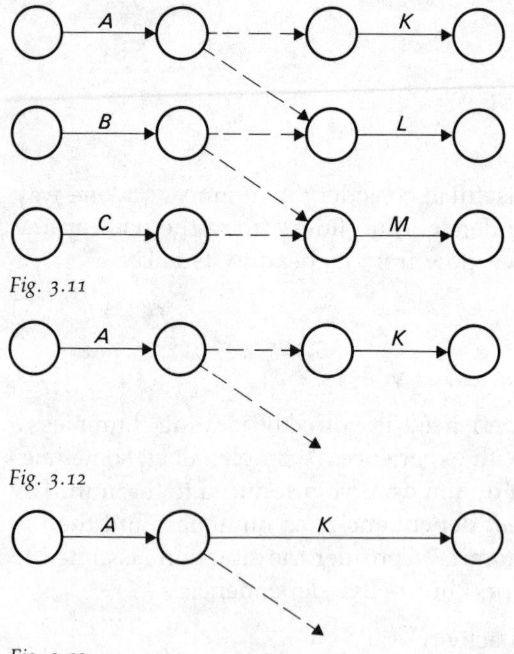

Fig. 3.11

Fig. 3.12

Fig. 3.13

However, neither the *A* to *L*, *B* to *L* nor *B* to *M* dummies can be removed without altering the logic of the situation, and they must, therefore, remain. The *C* to *M* dummy can be incorporated into *C* without any changes resulting, and the network will reduce to that given in Fig. 3.14.

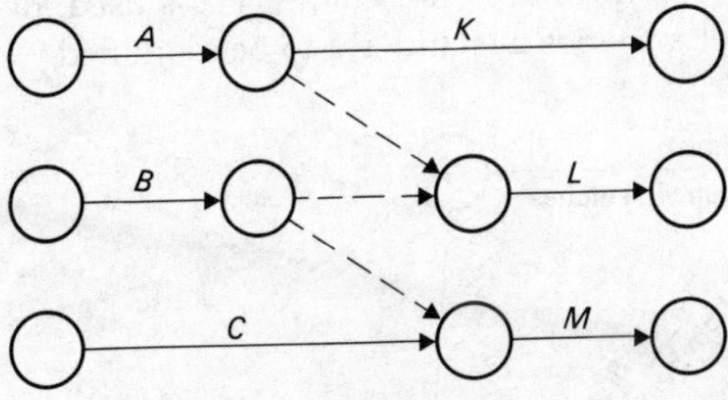

Fig. 3.14

No simple rule can be stated for use in this 'cleaning-up' process—the only test that can be applied is whether the absorption of a dummy changes the logic.

Note: This 'all dummy' approach in activity-on-arrow (AoA) initially produces diagrams very similar to activity-on-node (AoN) diagrams.

Overlapping activities

In all that has been said so far, it has been assumed that activities are quite discrete, the start of a succeeding activity being delayed until a previous activity is complete. There are many occasions, however, when this is not so: a succeeding activity *can* start when the previous activity is only partly complete. For example, if a large number of drawings are being prepared, the final drawings being traced in ink from pencil sketches, it might be thought that the arrow diagram would be:

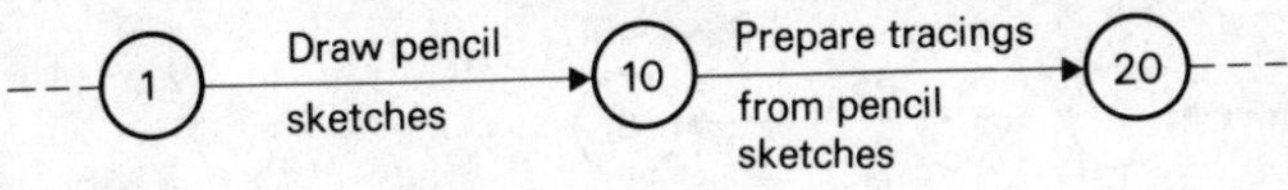

Fig. 3.15

but this would indicate that *no* tracings could be prepared until *all* pencil sketches were complete. It may well be that tracing can start after some pencil sketches are complete, and that thereafter sketching and tracing will go on concurrently, sketches being fed through as they are completed to the tracers who will eventually finish tracing some time after the last sketch is received. This can be represented by breaking both activities into two fractions (Fig. 3.16). This shows that it is not possible to complete tracings until all the sketches are complete and the first tracings have been finished. The dummy 3–4 is used, of course, to enable the parallel activities 2–3 and 2–4 to be identified.

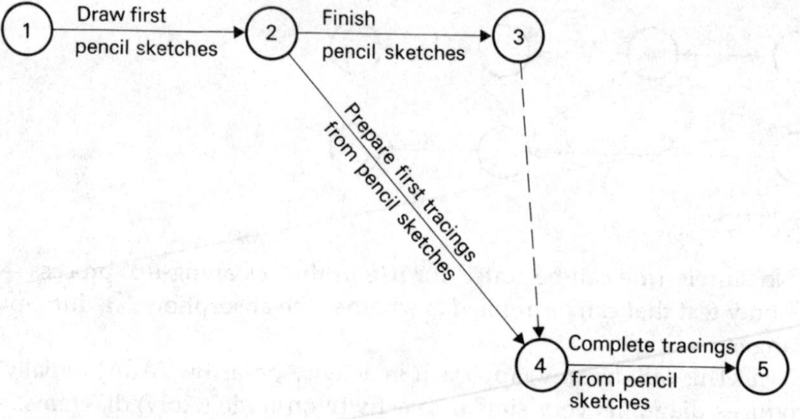

Fig. 3.16

This 'triangular' diagram represents the situation quite clearly but, when a number of these triangles are added in series, the drawing tends to become confused. A device that reduces the confusion is to draw the second activity arrow as a line 'bent' at right-angles. Thus, if the two activities are *P* and *R* 'broken' into *P*1 and *P*2, *R*1 and *R*2, the diagram will become as shown in Fig. 3.17 which is much clearer than the 'triangular' form, reinforcing the advice that activities should be drawn parallel to the axis of the diagram and hence parallel to each other.

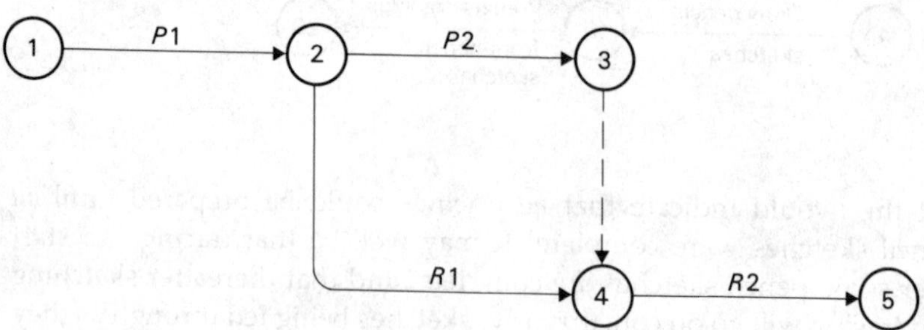

Fig. 3.17

If it is now desired to add another concurrent activity S, this can be done by 'breaking' $R1$ into two parts $R1/1$ and $R1/2$, thus permitting R to start part way through $S1$ (Fig. 3.18).

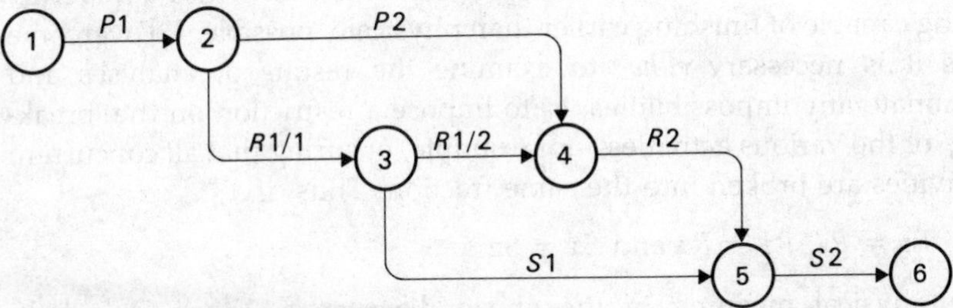

Fig. 3.18

Alternatively, it may be decided that $S1$ cannot start until *the whole* of $R1$ is completed (Fig. 3.19).

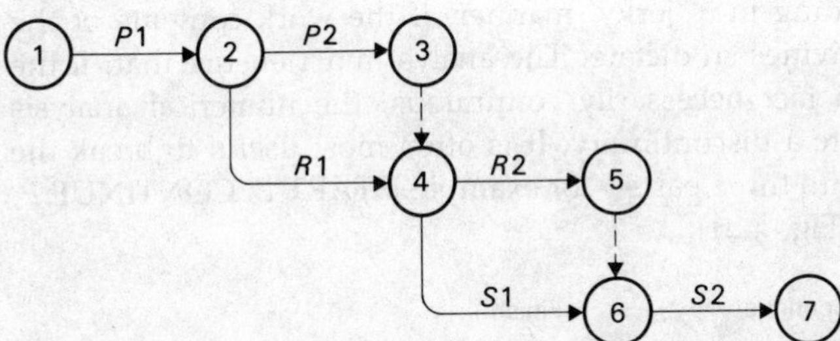

Fig. 3.19

Note: In Fig. 3.19, $S1$ depends on $P2$ through the dummy activity 3–4. If this is not so, and $S1$ can start with $R1$ complete and $P2$ incomplete, then a dummy between the junction of $R1$ and $S1$ and the junction of activity 3–4 and 2 will release the dependency. It will also remove the necessity for dummies 3–4 and 5–6 (Fig. 3.20).

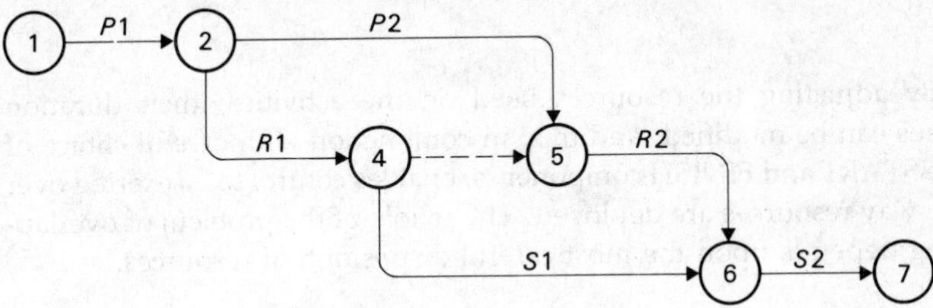

Fig. 3.20

All these diagrams represent similar situations, namely a resultant activity starting before its originating activity is complete. Care must be exercised in using and analyzing these networks, since some combinations of duration times can result in subsequent activities apparently being capable of finishing earlier than physically possible. To overcome this it is necessary *either* to examine the results of analyses and eliminate any impossibilities, *or* to impose a restriction on the 'breaking' of the various activities—for example, ensuring that all concurrent activities are broken into the same fraction. Thus, if

$$P1 = P2,\ R1 = R2 \text{ and } S1 = S2$$

the physical meaning of the above diagram is 'When P is half-completed, R is started, the *second* half of R not starting until the *second* half of P is completed. The *first* half of S is only started when the *first* half of R is completed, and the *second* half of S is only started when the second half of R is completed.' Of course, this may result (apparently) in tasks proceeding in a 'jerky' manner, if the work contents of the overlapping activities so dictate. The analyst must ensure that, if the activities are in fact necessarily continuous, the numerical analysis does not indicate a discontinuity. It is often most useful to break the base activities into three parts—for example, START P, CONTINUE P, and FINISH P (Fig. 3.21).

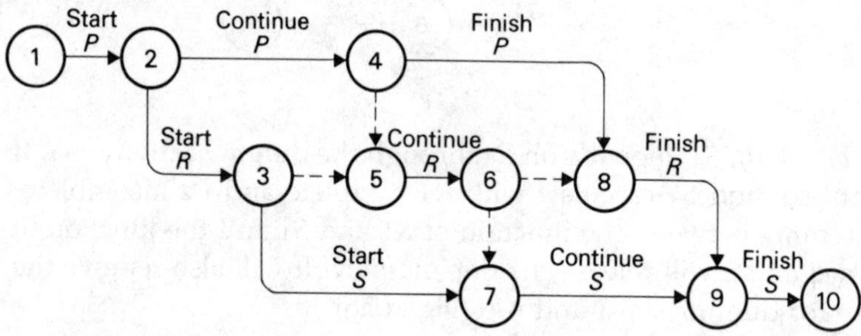

Fig. 3.21

By adjusting the resources used on the activities, their duration times can be modified, and this, in conjunction with careful choice of the START and FINISH components, enables control to be exerted over the way resources are deployed. The whole of the problem of overlapping depends upon the most careful disposition of resources.

4 Activity-on-node networking

The term 'activity-on-node' (AoN) networking implies not a single system of networking but a family. Of these AoN systems possibly the best known and most used is the *'method of potentials'* (MoP) by Mons. B. Roy, although there are indications that IBM's *'precedence diagramming'* is becoming popular. Since MoP is the simplest technique it will be described here in some detail: its family resemblance to the precedence diagram is so great that translation from one to the other is not difficult.

Elements of an AoN network

The network is made up of only two basic elements:

(1) An activity: an element of the work entailed in the project. This activity has precisely the same characteristics as the activity in AoA, namely that it is an element of the work entailed in the project. While resources may not be used ('wait for paint to dry'), an activity is essentially that which is necessary to the project. It is represented, however, by a *node*, usually drawn as a rectangle. Thus, an activity *A* would be shown as in Fig. 4.1. (Within the AoN family there are slight differences concerning the other information that is placed in the node. Reference will be made to these later.)

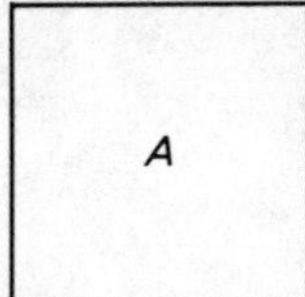

Fig. 4.1

(2) A dependency or sequence arrow that shows the interrelationships between various activities. Thus, if activity *B* depends upon (that is, *must* follow) activity *A* the diagram would be as given in Fig. 4.2.

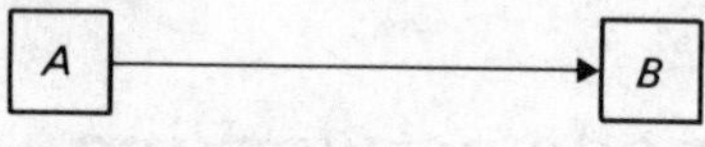

Fig. 4.2

These two symbols alone allow the representation of a project without the need for the dummy of AoA (see p. 19). Thus, if activity *K* depends on activity *A* and activity *L* depends on activities *A* and *B* the diagram would be as in Fig. 4.3.

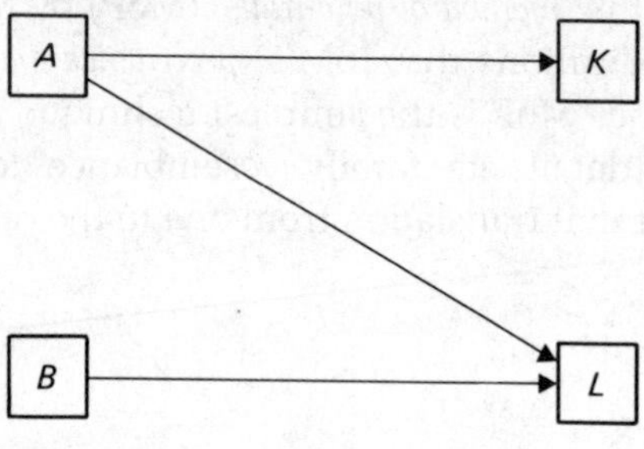

Fig. 4.3

Similarly:

> Activity *K* depends on activity *A*.
> Activity *L* depends on activities *A* and *B*.
> Activity *M* depends on activity *B*.

would give a diagram as in Fig. 4.4.

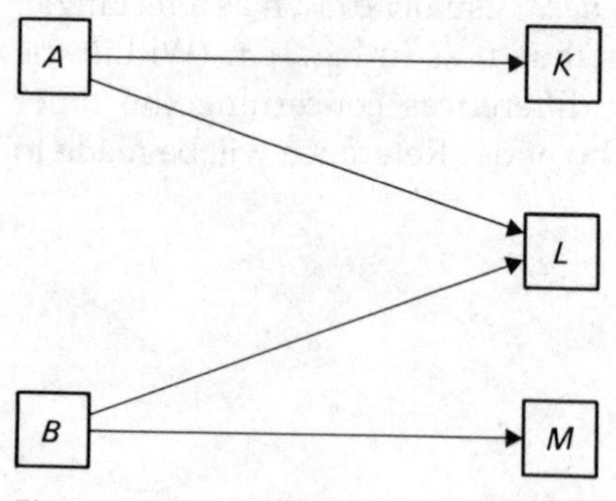

Fig. 4.4

While:

> Activity *K* depends on activity *A*.
> Activity *L* depends on activities *A* and *B*.
> Activity *M* depends on activities *B* and *C*.

would give a diagram as in Fig. 4.5. (c.f. pp. 21, 22 and 25.)

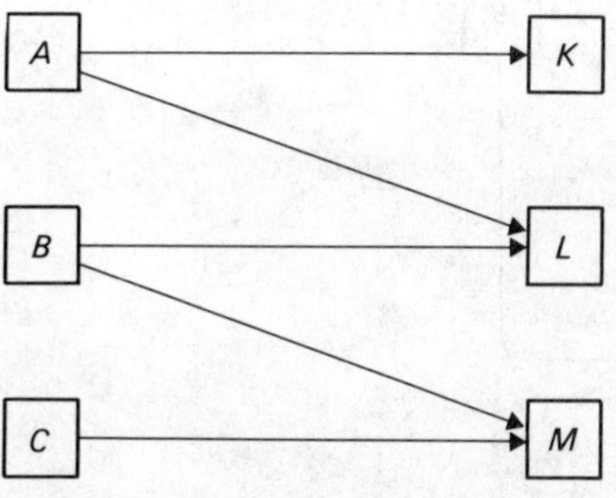

Fig. 4.5

Freedom from the need to introduce dummies is one of the most frequently cited advantages of AoN networking. While accepting this considerable benefit it must be pointed out that an AoN diagram is likely to be larger, and appear more complex, than the equivalent AoA network.

The representation of time in MoP

While CPA uses a subscript to the activity arrow to represent the duration time of the activity, MoP uses a subscript to the dependency arrow to denote the dependency time, that is, *the time that must elapse between the START of an activity and the START of the succeeding dependent activity*. This permits considerable flexibility in showing the time relationships between activities and constitutes one advantage of MoP over CPA. Since the dependency time is not necessarily the duration time of the tail activity, the duration time may be 'lost', and for this reason the author prefers to include the duration time within the activity node.

Negative constraints in MoP

The above dependency time may be said to form a *positive* constraint upon the start of a succeeding activity. A different sort of constraint, the so-called *negative* constraint may be incorporated in the diagram. Thus, if the interval between the completion of *A* and the start of *B* may not exceed *X* the diagram would appear as in Fig. 4.6. The value of this negative constraint in practice is more apparent than real.

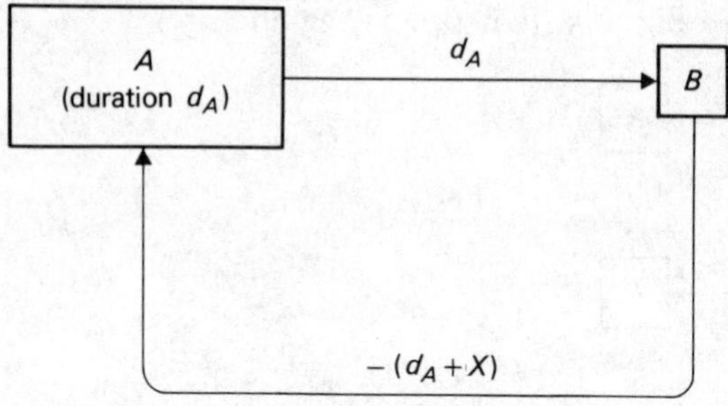

Fig. 4.6

Activity duration times

In AoN it is usual to include the activity duration time *within* the node. Thus, if activity *A* requires 16 units of time for its completion, then it will be represented as in Fig. 4.7.

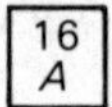

Fig. 4.7

If now activity *J* can only start when activity *A* is complete, then the diagram becomes as given in Fig. 4.8.

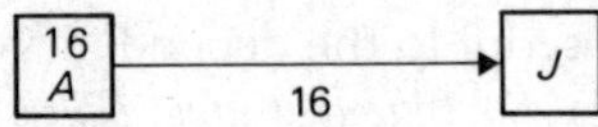

Fig. 4.8

If, however, activity *J* can start 12 units of time after the start of activity *A*, then it will become as in Fig. 4.9.

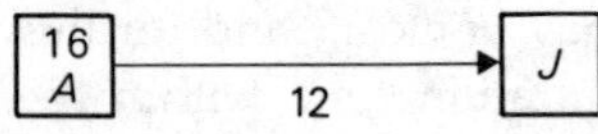

Fig. 4.9

Care must be taken in the use of dependency time to ensure that an unintentional absurdity does not result. For example, activity *A*, duration time 10 may precede activity *B*, duration time 5, and activity *B* may be capable of being started 1 unit of time after activity *A*. This might seem to give a diagram as in Fig. 4.10.

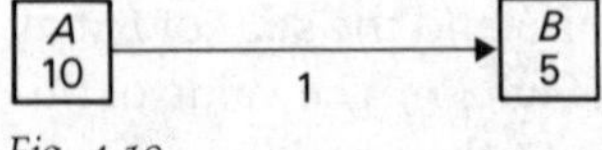

Fig. 4.10

This would give the situation:

	Start	*Finish*
Activity *A*	0	10
Activity *B*	1	6

that is, activity *B* would finish *before* its predecessor. This *may* satisfy the logic of the situation in which case the representation is valid. However, it could be that activity *A* must be complete before activity *B* can finish, in which case a false statement is being made.

Overlapping activities

A common situation is that where several activities 'overlap'. For example, assume there are three activities *P*, *R* and *S*. The diagram (Fig. 4.11)

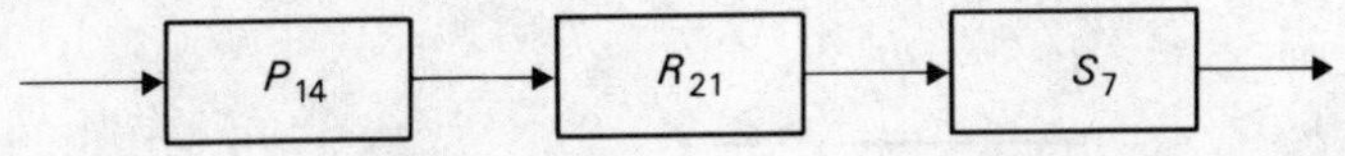

Fig. 4.11

would imply that the *whole* of *P* was complete before any of *R* were started, and that the *whole* of *R* was complete before any of *S* were started. Breaking each activity into three components 'start', 'continue' and 'finish' (Fig. 4.12)

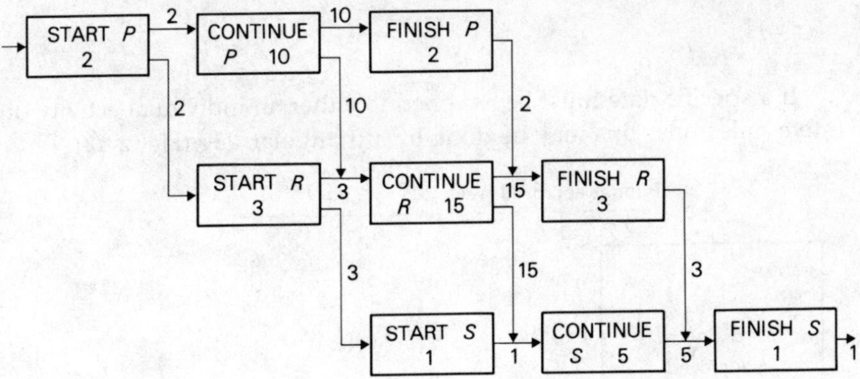

Fig. 4.12

would allow overlapping to be represented. This may result in tasks proceeding in a 'jerky' fashion if the work content of the overlapping activities so dictates. In practice, either this jerkiness must be tolerated—by allowing work rates to change and accommodate delays—or resources must be adjusted to give 'smooth' working.

Multi-dependency AoN presents a different solution to this problem.

Milestones in MoP

It is sometimes convenient in the life of a project to identify 'milestones' when particular decisions have to be taken or situations reviewed. Milestones usually represent the completion of a number of activities, and in CPA they are represented by events (or nodes). In MoP, events do not exist as such: however, it is possible to represent a milestone in MoP by using a fictitious activity, of duration 0 (Fig. 4.13).

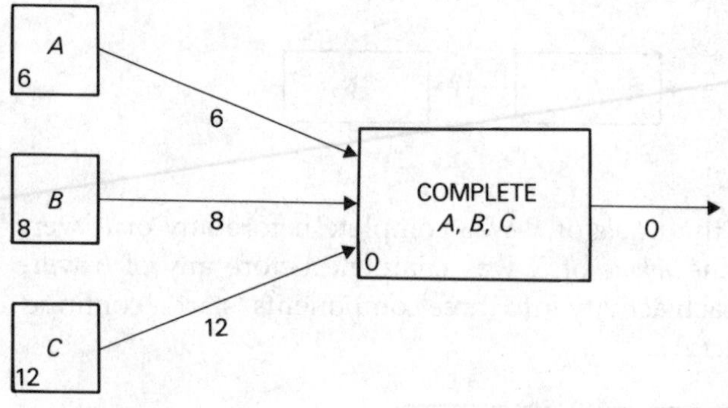

Fig. 4.13

If a specific date must be assigned to either an individual activity or to a milestone, this may be done by a triangular flag (Fig. 4.14).

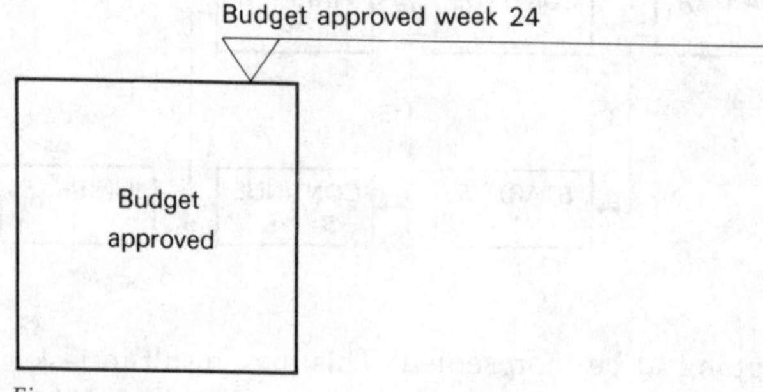

Fig. 4.14

The collector node

Occasionally, a very confused situation can arise when one group of activities are all dependent on each of another group. For example:

Activity *D* is dependent upon activities *A*, *B* and *C*.
Activity *E* is dependent upon activities *A*, *B* and *C*.
Activity *F* is dependent upon activities *A*, *B* and *C*.

Assuming that no activity may start till its predecessor is complete, then the diagram would become as shown in Fig. 4.15.

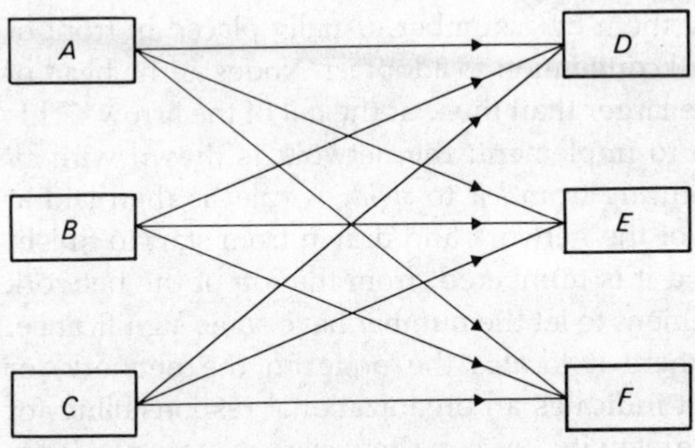

Fig. 4.15

A dummy activity (a 'collector' node) of zero duration would simplify this representation (Fig. 4.16).

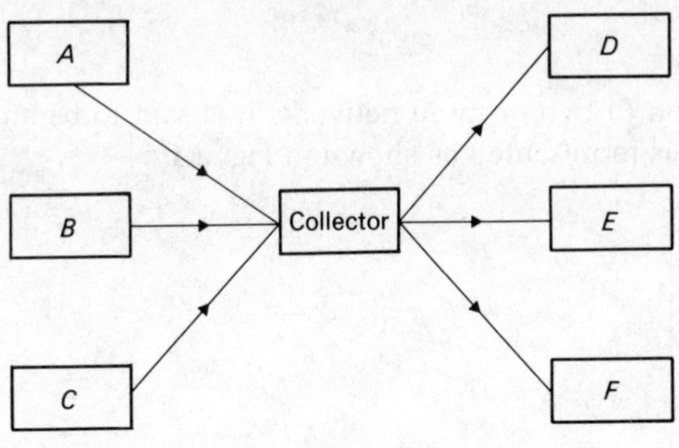

Fig. 4.16

Great care and discrimination must be employed in using a collector node as false dependencies can be created very easily.

Note: A collector node is not valid in the other main member of the

AoN family, the precedence diagram, since dependency arrows have a different significance according to the location of their heads/tails.

Start and finish nodes

In order to be able to carry out any calculations, all opening activities must emerge from a 'start' node of zero duration, and all closing activities must come together in a 'finish' node.

The identification of nodes

Nodes are identified by their description, but it is convenient and succinct also to identify them by a number, usually placed in front of the description. A useful convention to adopt is: 'Nodes at the head of a dependency arrow are larger than those at the tail of the arrow'. This is a simple convention to implement: the network is drawn with *all* dependency arrows pointing from *left* to *right*. A ruler is then laid at right-angles to the axis of the network and drawn from start to finish. As each node is exposed it is numbered, from the top of the network downwards. It is convenient to let the number have some significance, either geographical, where it locates the node on the network, or organizational, where it indicates an organizational responsibility for the carrying out of the activity. Avoid, however, excessively long, cumbersome nodes which are inconvenient to handle. It is also unwise to number consecutively (1, 2, 3 . . .) since a later insertion of a node may necessitate considerable renumbering. The author tends to number in fives (5, 10, 15 . . .).

Interfacing

If an activity is common to two or more networks it is said to be an 'interface' activity and is represented as shown in Fig. 4.17.

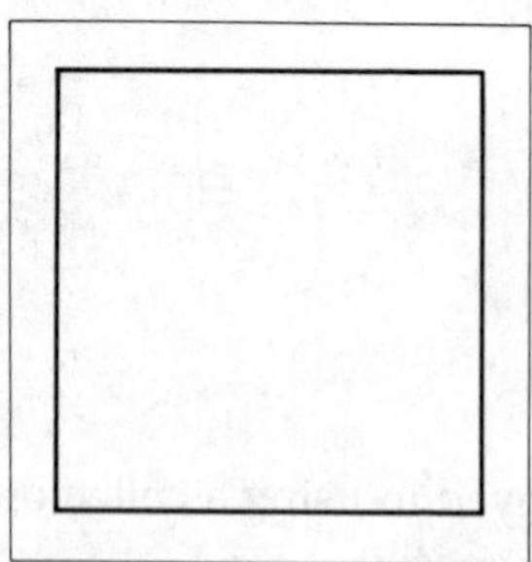

Fig. 4.17

Two errors in logic

Two errors in logic—*looping* and *dangling*—can build up in a network, particularly if it is a complex one.

Looping

Consideration will show that the loop in Fig. 4.18 must not occur, since this would represent an impossible situation: 'activity *R* depends on activity *Q* which depends on activity *P* which depends on activity *R*. . . .' Should such a loop apparently arise, either an arrow-head has been misplaced or an error has been incorporated in the logic (see p. 16).

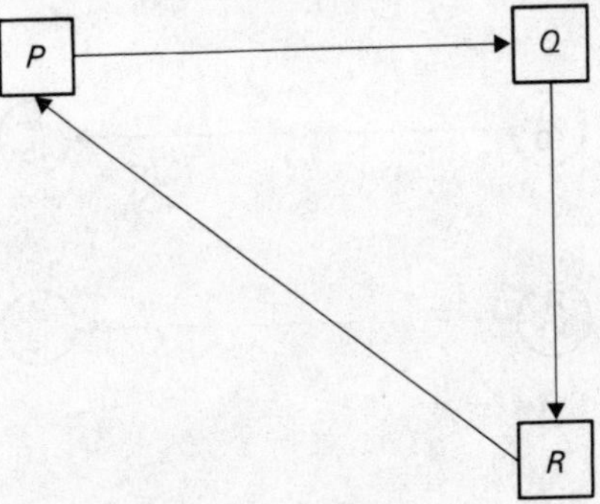

Fig. 4.18

Dangling

Similarly the situation represented by Fig. 4.19 is also at fault since activity *M* is undertaken with no result. Such dangling nodes often arise from hastily inserted afterthoughts.

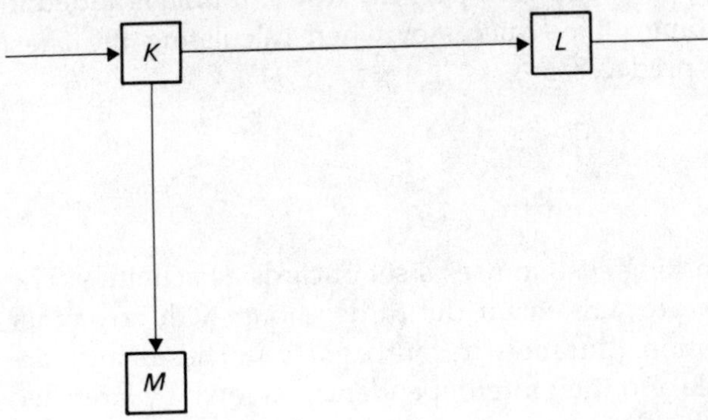

Fig. 4.19

The circle notation

An alternative to the MoP system is sometimes called the 'circle' notation (see Schaffer, Ritter and Meyer—Selected Reading) although present-day workers tend to enclose the activity in a rectangle rather than a circle.

In this system the activity is enclosed within the node circle, along with the activity duration, and these duration times are used in carrying out the forward and backward passes (unlike MoP as discussed above where they are effectively part of the activity description). The subscripts placed on the depending arrows indicate the degree of overlap, thus:

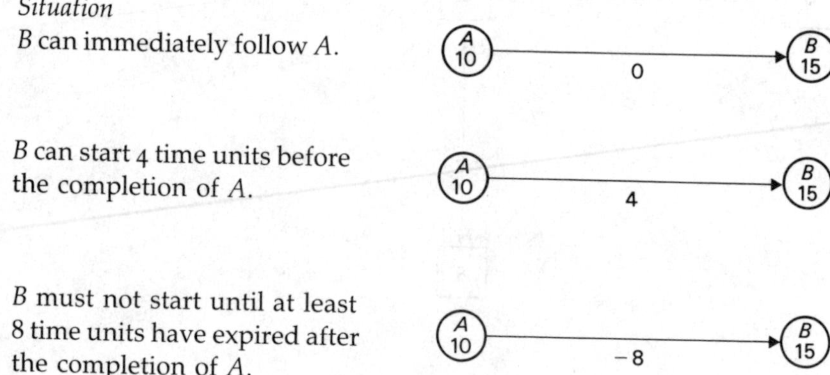

In the forward pass the restraint (or overlap) time is subtracted from the duration of the predecessor in calculating the earlier starting time of a successor, and in the backward pass the restraint time is added to the latest starting time of the successor when calculating the latest starting time of the predecessor.

Card networking

An ingenious networking system uses a set of cards as activities. The activity descriptions are written on the cards, along with any other appropriate information (duration, resources, etc.). These cards are then laid upon a table and their interdependencies shown by stretched strings or tapes. By this means the whole process of drawing a network is accelerated, and the originator claims that participation is increased. This system is completely described in 'Speeded Methods of Network Planning', by J. A. Larkin (*The Production Engineer*, February, 1968).

5 Drawing the network

The network as a statement of policy

Policy is sometimes defined as the means whereby an objective is to be achieved, and in this sense a network can be assumed to be a formal and explicit statement of policy. This concept will be found useful when considering the amount of detail that should be displayed.

In general, as one descends the hierarchy of an organization, the detail given in a policy statement increases, while its scope decreases. For example, the top management may decide to produce a new product *X*, and within the overall policy of the organization this might appear as shown in Fig. 5.1.

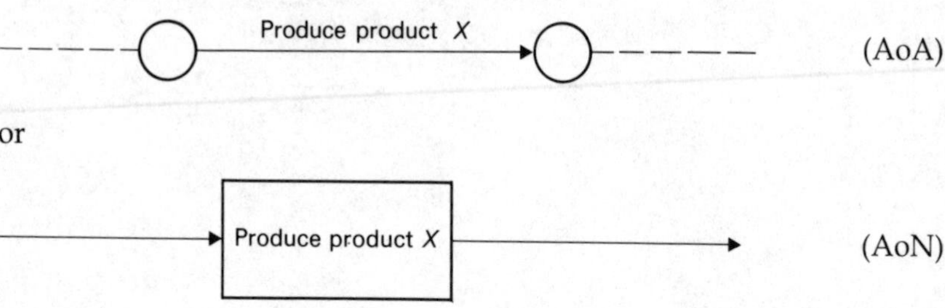

Fig. 5.1

The next level of management might be concerned with the deployment of the design and production resources, and the single arrow might become as in Fig. 5.2.

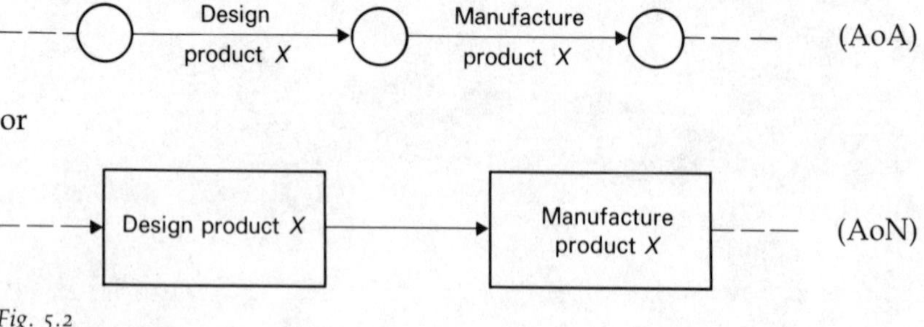

Fig. 5.2

The design manager at the next level might then translate the single 'Design Product *X*' arrow as a more complex network, of which a part might be as shown in Fig. 5.3.

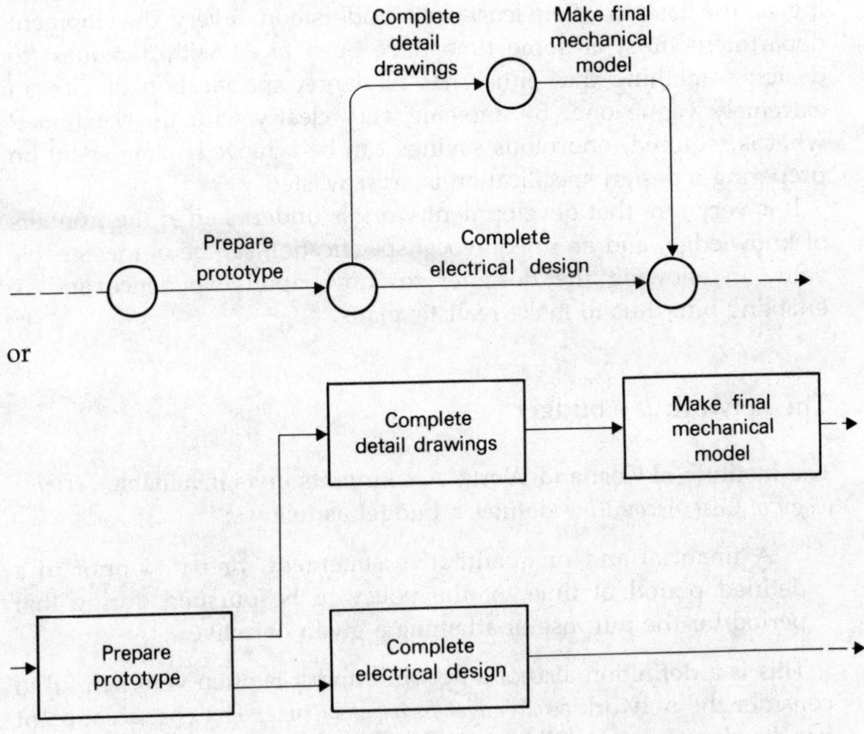

Fig. 5.3

The chief electrical engineer might then distribute the electrical designing among a number of different design engineers, so that in turn a single arrow at one level becomes a network at a lower level.

This idea of a network as a statement of policy is useful, and its implications are substantial. Policy denotes *known* objectives; it is not possible to play down sensible policy unless the purpose for which it is in being is clearly known and explicitly stated.

The first action, therefore, when drawing a network is to *define the purpose* of the project being considered. Stated thus baldly, the statement appears trite and, indeed, in some cases, the objectives are very clearly known. Even in construction and manufacturing however, where 'purpose' may appear obvious, some thought on this subject is valuable. A 'feasibility' network is likely to be very different to an operating network, a 'resource' network *may* be different to a time network and so on.

It is in design and development, an area where project network techniques (PNT) can be very usefully employed, that the understanding of the task is often least well understood. Every development department must at some time have been faced with a request to design something that either has no target specification or else an extremely vague one. By agreeing very clearly with the 'customer' what is required, enormous savings can be achieved. Time spent on preparing a design specification is *never* wasted.

It is very rare that development work is undertaken at the frontiers of knowledge, and an unequivocal specification can be of inestimable value in allowing the designer to draw upon experience and in enabling him thus to make realistic plans.

The network as a budget

The Institute of Cost and Works Accountants in its invaluable *Terminology of Cost Accounting* defines a budget as follows:

> 'A financial and/or quantitative statement, prepared prior to a defined period of time, of the policy to be pursued during that period for the purpose of attaining a given objective.'

This is a definition also of a network and it is often very helpful to consider the network as *a budget in terms of time*. The cost accountant has developed much skill in the assembly and use of financial budgets, and it is prudent to consider this experience when drawing and using networks. The author has found this parallel particularly valuable *when considering the detail that should be incorporated into a network*. A hierarchy of networks is just as appropriate as a hierarchy of budgets: it will enable problems to be identified without a mass of unnecessary detail, and it can locate responsibility at an appropriate point in the structure of the company.

Drawing the network

When drawing an arrow diagram of a project, the major activities are fairly readily identified, and these should be approximately located in their correct positions relative to each other and the start and finish events on a large sheet of paper. While it is possible at this stage to prepare a list of activity descriptions, it is always more convenient to write the descriptions on the diagram itself. For this reason it is unwise to try to make the arrows or nodes too small. Descriptions may well be abbreviated ('fdns' for 'foundations', 'cmpnts' for 'components' . . .) and some organizations set up a glossary of approved abbreviations.

A list of activities, amplifying the descriptions and specifying times and resources used is often useful, but this is probably best prepared once the network has been tidied up and tested. Some writers recommend that the first task to be undertaken is to prepare an activity list and from this prepare the diagram. The author has found this practice to be inhibiting. The list once prepared takes on a rigidity from which it is difficult to depart, and a feeling is inculcated that once all the activities in the list are 'fixed' into the network the planning is complete. It is often the act of preparing the network and debating the logic which clarifies the activities.

It is also unnecessary at this stage to try to make arrows straight, or always moving from left to right. Work on the diagram can proceed from both the start and the finish, and it is sometimes found that the project divides itself into a series of inter-related chains, and completing one chain at a time can be very helpful. The most useful pieces of equipment at this stage are a pencil and a good eraser: chalk and a blackboard are excellent alternatives.

The major activities having been drawn, the network can be completed by filling in the minor activities. A problem that repeatedly arises is to decide when to stop in writing down minor activities. If too much detail is written into a diagram it becomes excessively large, and the subsequent analysis increases in complexity without usefully increasing in value. There is, furthermore, the danger of losing sight of the physical realities underlying the diagram, and the whole analysis declining into a mathematical exercise. Rules on this do not seem to emerge, but it is suggested that three lines of inquiry can usefully be pursued:

(1) Can separate resources be shown with separate arrows?
(2) Does any single arrow cover responsibilities assignable to more than one person?
(3) Is the detail greater than that which is necessary for the person employing the network to make sensible decisions?

Almost certainly too little detail is preferable to too much.

It must be remembered, of course, that the definition of a resource, and the accountability for a responsibility, will apparently change with the level at which a plan is being made.

The first rough diagram may now be re-drawn, and the straightening-up and disposition of arrows checked in accordance with the conventions described above. It should be realized that, as arrows are not vectors, their lengths and orientations are determined *only* by the convenience of drawing and the logic behind the project. The properties of activities must be written into the diagram, but the location of an activity, by considering that which takes place *previously, concurrently* and *subsequently,* will also be found very helpful. It is also wise to investigate the need for activity-on-arrow (AoA) dummies in ensuring that where necessary they are inserted.

The author has found the following practice invaluable: once a network has been drawn, start from the final node and move up each activity, asking the question:

> What had to be done before this activity could take place?

Having reached the first node, move down each activity asking the question:

> What can now be done after the completion of this activity?

Carefully and systematically carried out, this procedure will be found to be of great assistance in ensuring logical cohesion.

As pointed out earlier the network is a statement of policy and, consequently, once the network is adopted, it commits the organization to a course of action, along with all the concomitant administrative procedures. Numbers of people—for example, departmental managers, site foremen, section leaders—are thus committed to *and will be held responsible for* carrying out tasks laid down in the network. If for no other reason, therefore, it is very wise indeed for the planner drawing the network to enlist the aid of the appropriate executives when drawing each part of the diagram. This may mean that the planner has to sit down separately and in conference with all the interested parties while individual responsibilities are being painfully worked out and agreed. It is often a temptation for a planner to try to carry out the whole operation by himself. This temptation should be resisted except in the case of very simple or frequently repeated tasks.

In order to locate responsibility and authority quite unambiguously, it is helpful to redraw the diagram so that it is divided horizontally into responsibility areas, and vertically into broad time areas. This will result in a diagram for an AoA network appearing as shown in Fig. 5.4

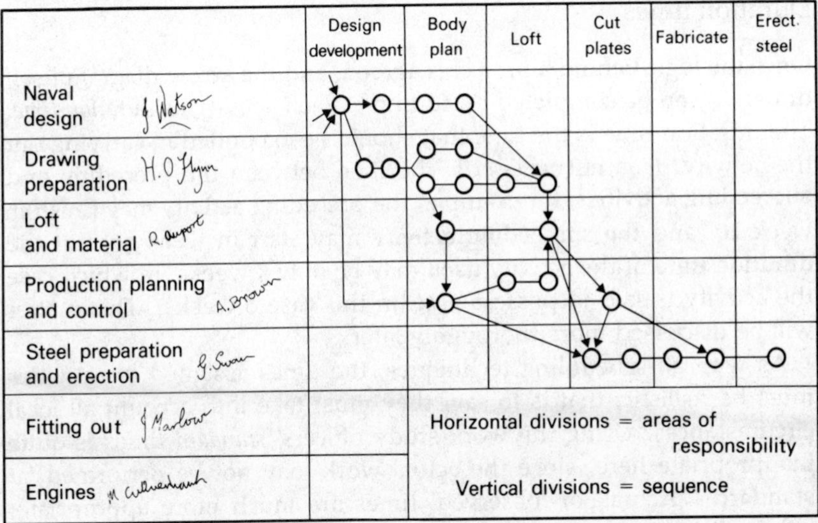

(By permission of International Computers & Tabulators Ltd.)

Fig. 5.4 Refinements to a network

to ensure that responsibilities are understood *and accepted*, the signatures of the appropriate officers are required to appear on the diagram itself. This type of technique can prove of considerable benefit to the structure of the company, and some of the results obtained will be:

(1) A clear understanding by all managers of the work they are committed to do.
(2) The delineation of responsibilities between managers.
(3) An investigation into the organization of, and procedures used in, the company.
(4) The application of current experiences to the planning function.

Interfacing

When projects are particularly large or complex, it is sometimes desirable to construct a number of small networks based either upon resources or responsibilities. These can then be amalgamated into a larger complete network by means of the common activities. These are conventionally represented either by double concentric nodes or double parallel lines with a single arrow-head, and they are known as *interface* elements, the amalgamating procedure being *interfacing*.

Duration times

Once the logic behind a project is agreed, and the arrow diagram itself drawn, it can be completed by adding to each activity its *duration time*. The duration time is the time that should be expended in carrying out the activity. It is not *necessarily* the time between the preceding and succeeding activities; for example, the preceding activity may finish in week 10, and the succeeding activity may start in week 20, but the duration time of the activity itself may be only 4 weeks, in which case the activity is said to possess *float* (in this case 6 weeks). This matter will be discussed more thoroughly later.

As with all scheduling techniques, the times assigned to activities must be realistic; that is to say, they must take into account all local circumstances. Using the work study officers' *standard* times is quite inappropriate here, since the actual work may not be performed 'at standard'. 'Actual' or 'observed' times are much more appropriate, although in many cases such times are not available. However, the principle is clear: the duration times need to be *realistic* rather than *desirable*, and they should be accepted by those held responsible for their achievement. Again, the experience of the cost accountant is pertinent: costs should be agreed with, and not imposed upon, the manager concerned. The *Cost Handbook* (Ronald Press Company, New York) contains the following (section 20.20):

> 'Primary responsibility for preparation of the budgets should rest with the supervisors of the various segments of the business. For example . . . the sales manager should participate actively in the development of the sales budget, since he will be the individual primarily responsible for the execution of the sales plan. This general procedure is equally applicable to every other segment of the business and should be vigorously pursued. . . .
>
> 'In this connection Francis (*Controller*, vol. 22) points out that: "Budgets are frequently developed by one or two key individuals . . . sometimes management without the prior knowledge or approval of the operating executives in the (various) departments. This is the worst type of budget procedure and quickly defeats the objectives of forward planning."'

Assigning duration times

When assigning duration times throughout a project for an action that should be carried out as late as possible in the planning sequence, it is sometimes helpful to consider activities in a random sequence. The point here is that if, say, activities are considered sequentially in chains, it is possible that the duration time assigned to one activity might affect the choice of duration time for later activities. For example, if it is realized that a 'long' activity might jeopardize the overall completion date, there is a temptation to 'shrink' subsequent duration times to give an acceptable overall answer. As with any planning method, PNT is no more accurate than the information fed into it, and it is often very difficult to avoid unwittingly colouring estimates when the final answer can apparently be seen.

Duration times under uncertainty

Not infrequently, duration times are held to be impossible to estimate ('I've never designed one of these before . . .' 'We won't know what we have to do until we've taken it down . . .'). However, it is extremely rare to find a situation where it is not vital to carry out a task in a limited time: a design *must* be completed in such-and-such a time in order to allow the product to be sent to a customer/shown at an exhibition/submitted to test conditions/. . ., and equipment usually has to be repaired before a particular date in order to keep production going/allow the chairman to go on his holidays/avoid a power cut/. . . and so on.

In situations like these it may be useful to proceed as follows:

(1) Establish the *purpose* of the task as accurately, and in as much detail, as possible.
(2) Establish any fixed dates. If no completion date is given, agree a reasonable target completion date.
(3) Assign duration times wherever possible.
(4) Break 'uncertain' activities into smaller parts, for some of which times can be readily agreed since historical data are available.
(5) Examine the residual uncertain areas to see if there are any precedents that can act as guides.
(6) Finally, assign to these uncertain activities as much time as possible without over-running the agreed or imposed target date. These activities can then be re-examined, asking the question:

'Can activity *X* be completed in time *t*?'

which is a more pointed and stimulating question than the original:

'How long will it take to complete activity *X*?'

Remember that activities may have float, so that a high order of accuracy in estimating times may be unnecessary. Thus, an activity with a duration time of the order of 2 weeks may have a float of 20 weeks—it is hardly necessary to refine the 2-week estimate. Always separate the decisions on logic from the decisions on time and resources.

Not only must the times be realistic in the above terms, they must be set assuming:

(1) Normal or usual methods are employed.
(2) Normal resources are immediately available.
(3) No other demands are going to be made on the resources.

Once duration times are assigned, an analysis will follow and from the results of this analysis the above assumptions may have to be amended. To *start* assigning times assuming constraints frequently leads to a distorted network. The author recalls one case when a company asking for advice had said that 'carpenters are always critical'. On examining networks this was found to be the case, but probing revealed that *before the network was drawn* the planner 'knew' that the carpenters would be critical and this so distorted the logic of the network that the initial assumption was proved.

PERT

In highly uncertain areas it may be better to use a 'bracket' of times, giving an estimate of the 'best' and 'worst' and 'most likely'. This is the situation for which PERT was originally designed.

PERT operates by assuming that the three-time estimates form part of a population obeying a β-distribution, so that the 'best' (a), 'worst' (b) and 'most likely' (m) times can be compounded to give a single 'expected time' (t_e) as follows:

$$t_e = \frac{a + b + 4m}{6}$$

and this t_e is the time used in all calculations. The choice of the β-distribution is not justifiable on experimental grounds, but it is computationally easy to handle, and its users state that it gives significantly useful answers.

A further use of the three-time estimate technique is to calculate the probability of any event occurring at any particular time. This manipulation has been very extensively discussed in the original PERT publications referred to in the selected readings. It is the author's experience that three-time estimates are very occasionally useful to derive a single time estimate, but that the effort of carrying out the probability calculations is not repaid in any way by the value obtained from the resultant probabilities. One highly respected colleague of the author asserts that the three-time estimate technique only encourages sloppiness in estimating.

Numbering the nodes in AoA

The final task in network drawing is to number the node. For AoA this should be done in accordance with the convention set out on p. 11. If the re-drawing of the network has, as recommended, caused all arrows to show time flowing from left to right, then numbering becomes quite simple. A straight edge is laid across the network, at right-angles to its axis, drawn across the network from start to finish, the nodes being numbered as they are exposed. For AoN, numbering should be as set out on p. 36.

Notes on drawing arrow diagrams

(1) Conventions to be used:
 (i) Time flows from left to right.
 (ii) Preceding nodes have a lower number than succeeding nodes.
(2) Identify the objective.
(3) Identify major activities.
(4) Locate major activities on a large sheet of paper.
(5) Draw the first diagram, joining major activities to minor activities.
(6) Check activity locations by asking:
 (i) What *has* happened?
 (ii) What *is* happening?
 (iii) What *will* happen?

(7) For every activity ask:
   (i) What had to be done before this?
   (ii) What can be done now?
(8) Check:
   (i) No looping.
   (ii) No dangling.
   (iii) All events are complete if all entering activities are complete.
(9) Redraw the diagram, numbering nodes in accordance with 1(*b*) above. Remember:
   (i) Arrow lengths and orientations are not significant.
   (ii) Events can be separated by dummy activities (in AoA).
(10) Avoid excessive detail. It is useful to ask 'Can any activities be amalgamated?'
(11) Essential equipment: pencil, eraser and large sheets of paper.

The unique network

Many workers, particularly when they first come to use PNT, have a belief that there can only be one network for a project. This, in fact, is not so: except in the very simplest situation there are always alternative methods of performing work so that different planners may very well produce different networks, each being apparently equally correct. The virtue of a network, however, is that it immediately enables alternative plans to be compared. Sooner or later, however, one new network must be produced but it must always be recognized that this will inevitably be a compromise of some sort.

Calculations on networks

The previous chapters have discussed how a network may be drawn, and the work so done is, in itself, invaluable. It has imposed a discipline upon the planners, forcing consideration of *what* has to be done, *when*, *by whom* and *in what time*. It has also provided a clear, unambiguous statement of policy that is readily understood by all potential users. Thus, if no further action were taken, considerable benefits would already have been derived. However, it is possible, by using only the very simplest arithmetic, to extract a considerable amount of extra information. For activity-on-arrow (AoA) networks this is covered in Chapters 6 and 7 while for activity-on-node (AoN) networks it is covered in Chapters 9 and 10.

6 Analyzing the activity-on-arrow network: I Isolating the critical path

Activity and event times

The total project time (*TPT*) is the shortest time in which the project can be completed, and this is determined by a sequence (or sequences) of activities known as the critical path (or paths). To calculate the *TPT*, carry out a *forward pass* whereby the *earliest starting times* (*EST*) for each activity is calculated. In the calculation it will sometimes be necessary to refer to the *earliest finishing time* (*EFT*) of an activity, given by:

Earliest finishing time = earliest starting time + duration

The critical path is then identified by carrying out a backward pass whereby the *latest finishing time* (*LFT*) of an activity and its associated *latest starting time* (*LST*) are calculated, given by:

Latest starting time = latest finishing time − duration

A node has also two times associated with it, its earliest time (the earliest event time—*EET*) and its latest time (the latest event time—*LET*). The meaning of these two times is clear once reference is made to the fundamental properties of events and activities (p. 14).

The *EET* for a node is the earliest time at which any activity emerging from that node can start. The *LET* for a node is the latest time by which all activities entering that node can finish.

Thus, assume that node 15 (Fig. 6.1)

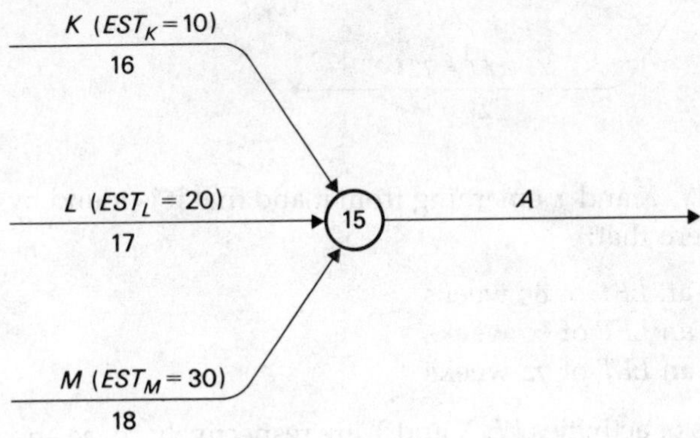

Fig. 6.1

has three activities, *K*, *L* and *M*, entering it, and that it is found by calculations elsewhere that:

> Activity *K* has an *EST* of 10 weeks
> Activity *L* has an *EST* of 20 weeks
> Activity *M* has an *EST* of 30 weeks

If the duration times of activities *K*, *L* and *M* are respectively 16, 17 and 18 weeks, then:

> Activity *K* has an *EFT* of (10 + 16) weeks = 26 weeks
> Activity *L* has an *EFT* of (20 + 17) weeks = 37 weeks
> Activity *M* has an *EFT* of (30 + 18) weeks = 48 weeks

The fundamental property of activities (p. 14) states:

> 'No activity may start until all previous activities in the same chain are complete.'

In the example being considered *all* the activities (*K*, *L* and *M*) preceding activity *A* are not complete until week 48—hence the *EST* for activity *A* is week 48, and the *EET* for event 15 is week 48.

In reverse, assume that node 121 (Fig. 6.2)

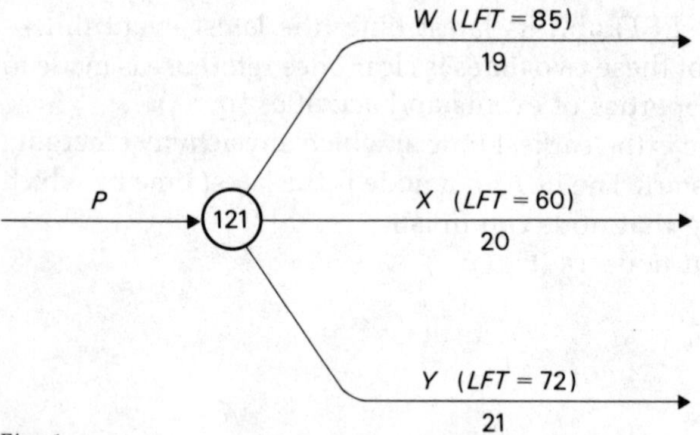

Fig. 6.2

has three activities *W*, *X* and *Y* emerging from it and that it is found by calculations elsewhere that:

> Activity *W* has an *LFT* of 85 weeks
> Activity *X* has an *LFT* of 60 weeks
> Activity *Y* has an *LFT* of 72 weeks

If the duration times of activities *W*, *X* and *Y* are respectively 19, 20 and 21 weeks, then:

Activity *W* has an *LST* of (85 – 19) weeks = 66 weeks
Activity *X* has an *LST* of (60 – 20) weeks = 40 weeks
Activity *Y* has an *LST* of (72 – 21) weeks = 51 weeks

The *LFT* of *P*, therefore, is that time which will permit *all* three activities *W*, *X* and *Y* to start as late as possible—that is week 40. The *LET* for event 121, therefore, is week 40.

The calculations in detail

The calculations will be carried out as shown in Fig. 6.3 (giving the time in weeks).

The forward pass

Assign to the first node, here node 1, an *EET*: 0, that is, activities *A*, *B* and *C* may start at the beginning of week 0. Then:

Node 2 has an *EET* of 16 weeks, that is, activity *J* may start at the beginning of week 16.
Node 3 has an *EET* of 20 weeks, that is, activities *E* and *D* may start at the beginning of week 20.
Node 7 has an *EET* of (20 + 15) weeks = 35 weeks, that is, activities *G* and *H* may start at the beginning of week 35.

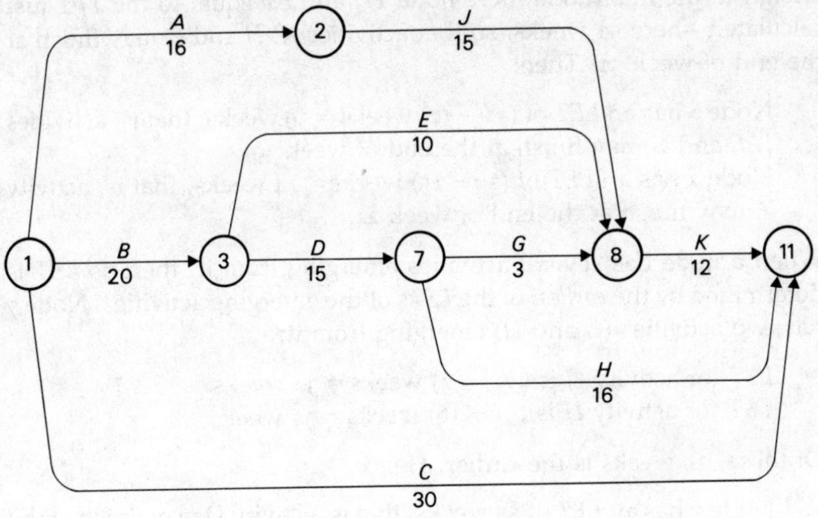

Fig. 6.3

When a node has several activities entering it, then its *EET* is determined by the *latest* of the *EFTs* of the entering activities. Node 8 has three activities (*J*, *E* and *G*) leading into it.

EFT for activity *J* is (16 + 15) weeks = 31 weeks
EFT for activity *E* is (20 + 10) weeks = 30 weeks
EFT for activity *G* is (35 + 3) weeks = 38 weeks

Of these, 38 weeks is the latest. Hence:

Node 8 has an *EET* of 38 weeks, that is, activity *K* may start at the beginning of week 38.

Similarly, for node 11:

EFT of activity *K* is (38 + 12) = 50 weeks
EFT of activity *H* is (35 + 16) = 51 weeks
EFT of activity *C* is (0 + 30) = 30 weeks

Of these finishing times, 51 weeks is the latest. The *EET* for node 11 is therefore 51 weeks. Since node 11 is a final node, this 51 weeks represents the minimum time in which the whole project can be completed—it is the *TPT*.

The backward pass

Assign to the final node, here node 11, an *LET* equal to the *TPT* just calculated—here 51 weeks, so that activities *K*, *H* and *C* may finish at the end of week 51. Then:

Node 8 has an *LET* of (51 − 12) weeks = 39 weeks, that is, activities *J*, *E* and *G* may finish at the end of week 39.
Node 2 has an *LET* of (39 − 15) weeks = 24 weeks, that is, activity *A* may finish at the end of week 24.

When a node has several activities emerging from it, then its *LET* is determined by the *earliest* of the *LSTs* of the emerging activities. Node 7 has two activities (*G* and *H*) emerging from it:

LST for activity *G* is (39 − 3) weeks = 36 weeks
LST for activity *H* is (51 − 16) weeks = 35 weeks

Of these, 35 weeks is the earlier. Hence:

Node 7 has an *LET* of 35 weeks, that is, activity *D* may finish at the end of week 35.

Similarly for node 3:

LST for activity *E* is (39 – 10) weeks = 29 weeks
LST for activity *D* is (35 – 15) weeks = 20 weeks

Hence:

Node 3 has an *LET* of 20 weeks, that is, activity *B* may finish at week 20.

and for node 1:

LST for activity *A* is (24 – 16) weeks = 8 weeks
LST for activity *B* is (20 – 20) weeks = 0 weeks
LST for activity C is (51 – 30) weeks = 21 weeks

Hence:

Node 1 has an *LET* of 0 weeks.

Since node 1 is the first node the *LST* of the *whole project* is week 0. This is to be expected: when the turn-round time at the final node is constant—that is, when the *EET* and *LET* of the final node are the same—the *EET* and *LET* of the first node must be the same since the forward pass adds up numbers that are then subtracted by the backward pass.

The calculations in practice

The above detailed calculations are given here for ease of explanation. In practice, it will invariably be found that, when not using a computer, the calculations are best carried out *on the diagram itself*. A number of techniques are available to do this: for example, marking the *EET* and *LET* with *E* = . . . *L* = . . . (Fig. 6.4).

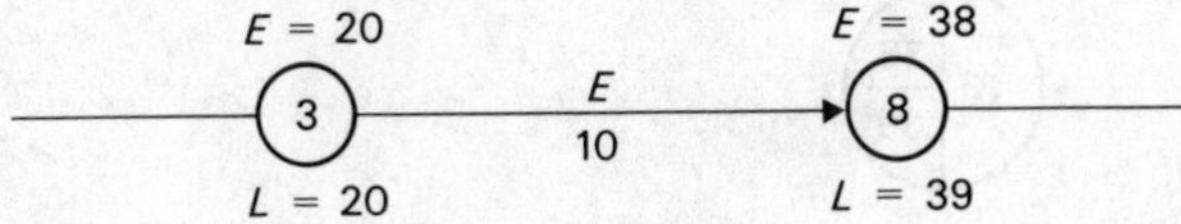

Fig. 6.4

or enclosing the *EET* and *LET* within geometrical symbols, for example, a square for the *EET* and a triangle for the *LET* (Fig. 6.5).

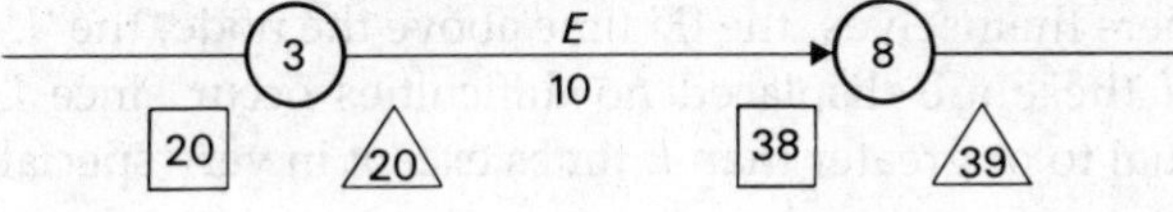

Fig. 6.5

An ingenious technique, attributable to D. Whattingham of CJB Ltd., is to replace the simple event circle by one divided into four segments, writing the event number in the bottom segment (Fig. 6.6).

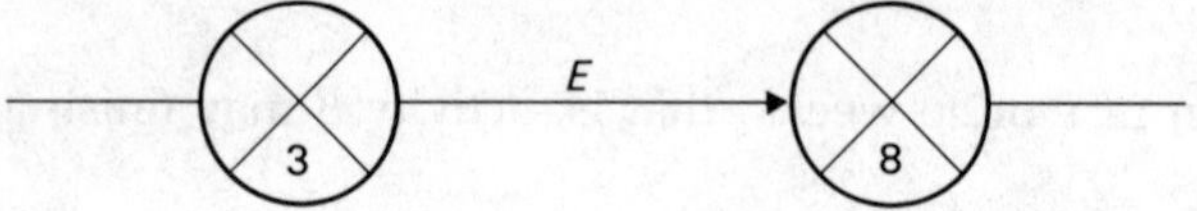

Fig. 6.6

The *EET* is then written in the left-hand segment, and the *LET* in the right-hand segment (Fig. 6.7).

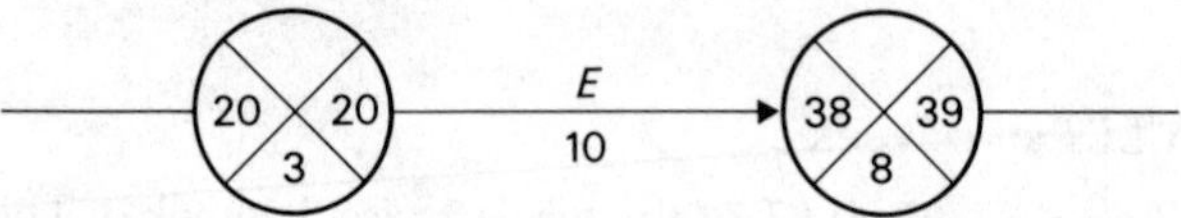

Fig. 6.7

This technique is particularly useful when recording progress once the project is under way, when the actual event time can be written into the top segment.

BS 6046: Part 2: 1981 recommends a trisected circle (Fig. 6.8).

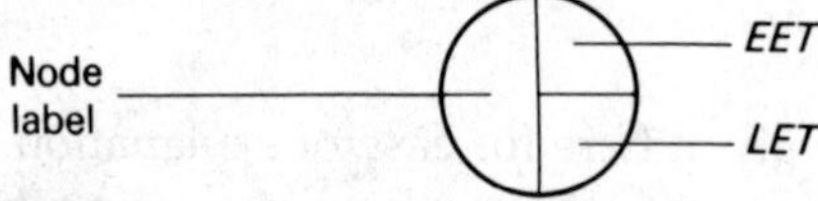

Fig. 6.8

Activity 3–8 would then appear as in Fig. 6.9.

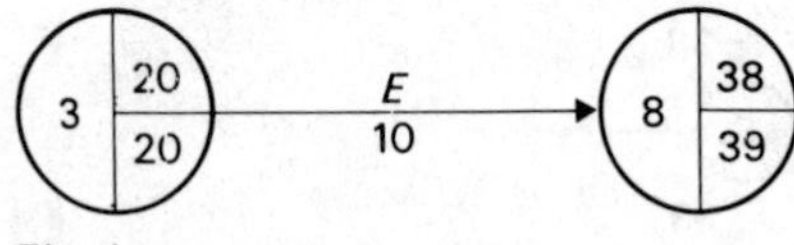

Fig. 6.9

The author finds that the divided circles tend to increase the size of the complete network, sometimes to unwieldy proportions. His own preference is either to use the $E = L =$ technique or, more frequently to write only the numbers themselves, the '*E*' time above the node, the '*L*' time below. Even if these are displaced no difficulties occur since *L* times are always equal to or greater than *E* times except in very special circumstances (Fig. 6.10).

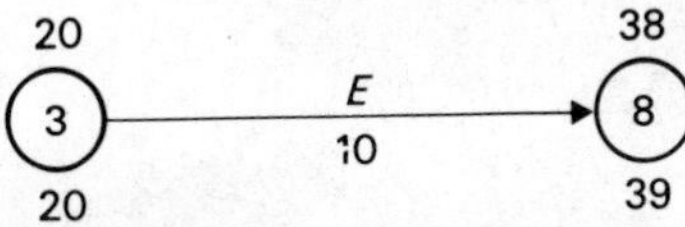

Fig. 6.10

The sample network using the $E = L =$ technique would be as in Fig. 6.11.

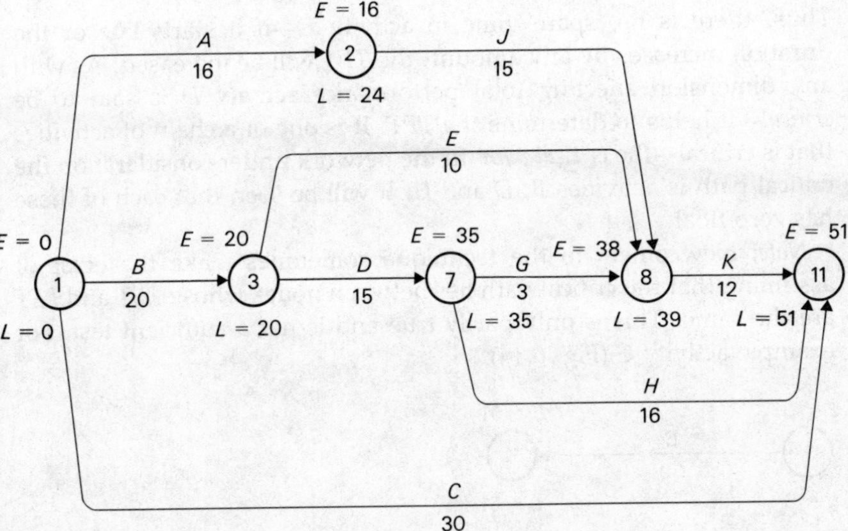

Fig. 6.11

The critical path

Consider activity *E*. The calculations show that the earliest time it *can* start—given by the *EET* of its tail node—is week 20, while the latest time it may finish is week 39. Thus, the time that can be made *available* for the activity is (39 − 20) weeks = 19 weeks. The time *required* for this activity—that is, its duration time—is 10 weeks. There are, therefore, (19 − 10) weeks = 9 weeks 'spare' ('float') for the performance of this activity. It may start 9 weeks late, finish 9 weeks early or occupy 9 weeks extra time without increasing the *TPT*.

Consider now activity *D* (Fig. 6.12).

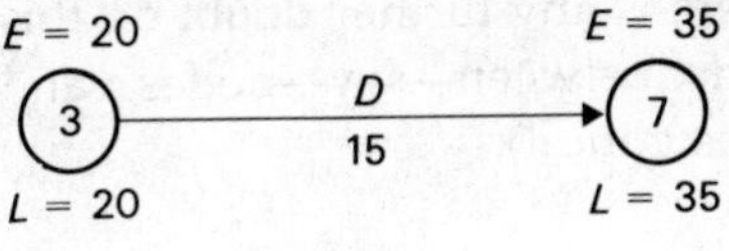

Fig. 6.12

LFT in weeks	= 35
EST in weeks	= 20
Time available in weeks	= 15
Time required in weeks	= 15
∴ Spare time in weeks	= 0

Thus, there is no 'spare' time in activity *D*—if it starts late or the duration increases by any amount, the *TPT* will be increased. As with any dimension affecting total performance activity *D* is said to be *critical*—it helps to determine the *TPT*. It is one of a chain of activities that is critical—the *critical path*. In the network under consideration the critical path is activities *B*, *D* and *H*. It will be seen that each of these has zero float.

Note: Newcomers to the technique sometimes make the error of assuming that the critical path lies between nodes whose *EET* and *LFT* are the same. This is only partly true and is not a sufficient test. For example activity C (Fig. 6.13)

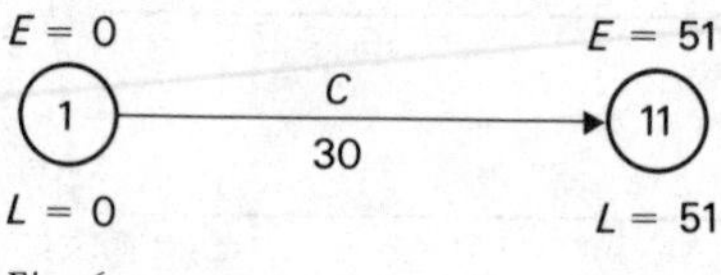

Fig. 6.13

lies between nodes whose earliest and latest times are the same. However:

LFT in weeks	= 51
EST in weeks	= 0
Time available in weeks	= 51
Time required in weeks	= 30
∴ Spare time in weeks	= 21

Clearly, activity C is not critical. (If there is any further doubt on this matter the reader should insert an activity between—say—nodes 1 and 7 with duration time of 1 and see if it is critical).

Float is the only means of identifying the critical path.

The critical path is conventionally indicated by a pair of small transverse parallel lines across the activities concerned (Fig. 6.14).

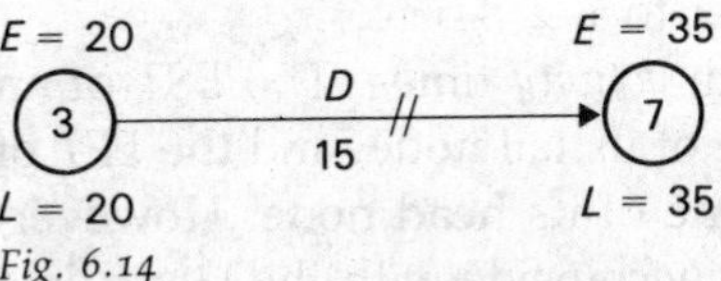

Fig. 6.14

Note: In practice, if a network is being calculated manually, it is rarely necessary to calculate float on all paths. Three simple factors characterize the critical path:

(1) It starts at the first node.
(2) It is continuous.
(3) It ends at the last node.

Activity times—a recapitulation

Since activities cannot start until their tail events are complete, and must not finish after their head events must start, the head and tail event times can be considered to fix boundaries between which activities can 'move'. It is possible to describe these 'movements' by four simple times:

(1) The *EST* is the earliest possible time at which an activity can start, and is given by the earliest time of the tail node. Thus, the *EST* for activity 2–8 is the earliest time for node 2—that is, 16 weeks.
(2) The *EFT* of an activity is the earliest possible time at which an activity can finish, and is given by adding the duration time to the *EST*; again, for activity 2–8 this is 16 + 15 weeks = 31 weeks.
(3) The *LFT* is found by taking the *LET* of the head node; again, for activity 2–8 this is the *LET* for node 8, that is 39 weeks.
(4) The *LST* is the latest possible time by which an activity can start, and is given by subtracting the duration time from the latest finish time; for activity 2–8 the *LST* is 39 − 15 weeks = 24 weeks.

Summarizing the above:

Activity			*Start time*		*Finish time*	
Number	*Description*	*Duration*	*Earliest*	*Latest*	*Earliest*	*Latest*
2–8	J	15	16	24	31	39

This can be done for all activities. The significance is that, considering activity 2–8, it must start between weeks 16 and 24 and must finish between weeks 31 and 39. An earlier start is impossible, while a later start will increase the overall performance time for the project; in fact, it will shift the critical path from 1–3–7–11 to 1–2–8–11.

Note: do not confuse *event* times with *activity* times. The *EST* of an activity coincides with the earliest time of its tail node, and the *LFT* of an activity coincides with the latest time of its head node. However, the *LST* of an activity does not *necessarily* coincide with the latest time of its associated tail event, nor does the *EFT* *necessarily* coincide with the earliest time of its head event; such coincidences only apply to activities on the critical path. Thus, the *LST* and the *EFT* in the table below *cannot* be read directly from the E = L = diagram, but *must* be derived from the *LFT* and the *EST*. It may be helpful to consider that event times *E* 'look forward' and event times *L* 'look backwards', so that in Fig. 6.15.

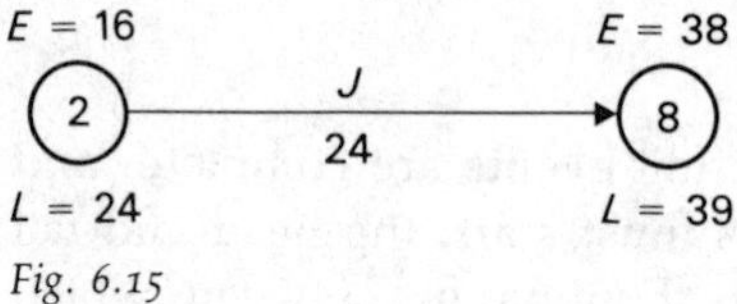

Fig. 6.15

E = 16 'looks forward' to provide one boundary for activity *J*, while E = 38 'looks forward' to provide a boundary for any activity emerging from activity 8. Equally *L* = 39 'looks backward' to provide a boundary for activity *J* while *L* = 24 'looks backward' to provide a boundary for any activity entering activity 2.

The earliest and latest start and finish times for the whole of the sample network are:

Activity			*Start time*		*Finish time*	
Number	*Description*	*Duration*	*Earliest*	*Latest*	*Earliest*	*Latest*
1– 2	*A*	16	0	8	16	24
1– 3	*B*	20	0	0	20	20
1–11	*C*	30	0	21	30	51
2– 8	*J*	15	16	24	31	39
3– 7	*D*	15	20	20	35	35
3– 8	*E*	10	20	29	30	39
7– 8	*G*	3	35	36	38	39
7–11	*H*	16	35	35	51	51
8–11	*K*	12	38	39	50	51

A matrix method of calculating earliest and latest event times

There are a number of alternative methods of calculation, one of which follows. However, the author feels that it is important that the physical meaning underlying earliest and latest times should be thoroughly understood, and that this understanding is best obtained by carrying out the calculations as set forward previously.

This alternative method uses a simple matrix, which is set up by drawing a square with two rows and two columns *more* than the number of events. Label the top left-hand corner *E* and the bottom left-hand corner *L*. From the left-hand corner of the second square in the first row, draw a diagonal through all the diagonal squares to the bottom right-hand corner of the last square in the penultimate row. Label the top half of the second square in the first row *j* and the bottom half *i*. Along the top row (i.e. opposite *j*) write all the event numbers, and down the second column (i.e. opposite *i*) also write all the event numbers. For the simple example network already discussed, the square will appear as in Fig. 6.16.

E	i \ j	1	2	3	7	8	11
	1						
	2						
	3						
	7						
	8						
	11						
L							

Fig. 6.16

Each small square is called a cell, and in the cell opposite head and tail numbers (i.e. opposite *i* and *j* numbers) fill in the duration time of the corresponding activity. For example, in the second row (labelled 1) under the fourth column (labelled 2) fill in the duration time of activity 1–2, that is, 16 weeks, and in the same row under the fifth column (labelled 3) fill in the duration time of activity 1–3, that is, 20 weeks, and in the same row in the last column (labelled 11) the duration time of activity 1–11. Repeat this for all activities. The square will then look like Fig. 6.17.

E	i \ j	1	2	3	7	8	11
	1		16	20			30
	2					15	
	3				15	10	
	7					3	16
	8						12
	11						
L							

Fig. 6.17

The square is now in a form from which calculations can be made. To find the *EÉT*s proceed as follows:

Start at the first event in the i column and move right along the row until the diagonal is reached (i.e. until $j = 1$ is reached). Set in the E column the figure found above the diagonal. In this case it is 0. (*Note*: this first step is unnecessary but is inserted to set the pattern for subsequent events.) Move now to the next event lower in the i column (in this case to $i = 2$) and again move to the diagonal. Set in the E column against $i = 2$ whatever number is found above the diagonal *plus* whatever is opposite that number in the E column—in this case $16 + 0 = 16$.

Proceed to $i = 3$ and $i = 7$ when $20 + 0$ and $15 + 20$ are entered in the E column. When $i = 8$, a slightly different situation arises since there are three numbers above the diagonal. The number entered in the E column is that number which is the greatest from the sum of the individual cells *plus* the corresponding E number. Thus, in this case there are three cells to be considered with values 15, 10 and 3 respectively, the corresponding E numbers being 16, 20 and 35, thus:

E	$i = 8$	$E + j$
16	15	31
20	10	30
35	3	[38]

The greatest sum is 38, and this is entered in the E column against $i = 8$. The same procedure is then repeated for $i = 11$, that is, the individual cells in the $j = 11$ column are added to the corresponding E column, and whichever sum is greatest is entered in the E column against $i = 11$, thus:

E	$i=11$	$E+j$
0	30	30
35	16	51
38	12	50

so that 51 is entered against $i=11$.

This whole process will fill the E column and give the earliest times for each event opposite the corresponding event number. To find the *LET* the whole process is reversed, the latest times appearing in the L row at the bottom, starting from the bottom right-hand corner.

The latest time for the final event is, as previously, the same as its earliest time, and this is filled in the bottom right-hand corner, that is, 51 is filled in in the L row in the $j=11$ column. Proceed to the next column and move upwards until the diagonal is reached, and then move along that row, subtracting the number in the row from the corresponding L number. In this case, there is only one number in the row (12) and this is subtracted from the 51 in its own column. Thus, in the $j=8$ column, a value of $L=51-12=39$ is entered.

The same procedure applies for $j=7$ except that in this case there are two numbers in the $i=7$ row, namely 3 and 16 with corresponding L numbers of 39 and 51, and the smaller difference is entered thus:

$i=7$	3	16
L	39	51
$L-i$	36	[35]

so that 35 is entered against $j=7$. Similarly, for $j=3$, proceed up the $j=3$ column until the diagonal is reached; then proceed along the intersecting $i=3$ row, subtracting the activity times from the previously recorded L times thus:

$i=3$	15	10
L	35	39
$L-i$	[20]	29

so that an L figure of 20 is entered beneath $j=3$. For $j=2$, only one figure is possible $(39-15)$, while for $j=1$, there are three possible figures:

$i=1$	16	20	30
L	24	20	51
$L-i$	8	[0]	21

and a figure of 0 is entered under the $j=1$ column. This provides a check on the whole calculation, since the earliest and latest times of the *first* event (like the final event) must be the same.

The completed matrix will appear as shown in Fig. 6.18; from this the various *EET*s and *LET*s can be directly read.

One useful by-product of this method is that it gives a simple way of identifying those events that are directly linked to others; thus from the first line one can see that event 1 is directly linked to events 2, 3 and 11. The second row shows that event 2 is only linked to event 8, and so on. In a small network this identification is of little value, but in larger networks it can save much searching.

E	*i* \ *j*	1	2	3	7	8	11
0	1		16	20			30
16	2					15	
20	3				15	10	
35	7					3	16
38	8						12
51	11						
L		0	24	20	35	39	51

Fig. 6.18

7 Analyzing the activity-on-arrow network: II Float or slack

As has already been described, the earliest tail and the latest head event times form boundaries within which activities are able to move.

Total float

Looking again at activity *J*, as shown in Fig. 7.1, it will be seen that

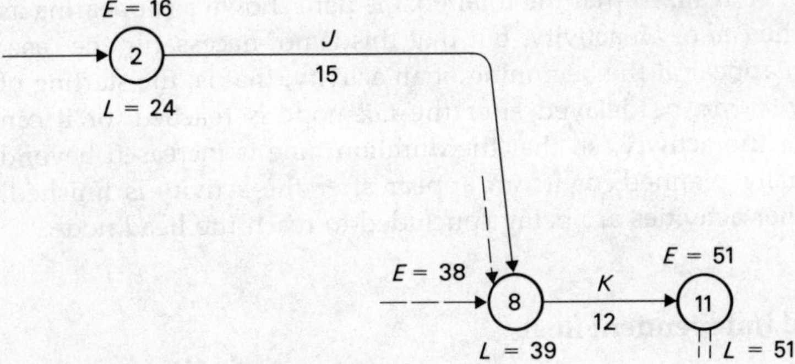

Fig. 7.1 Refer back to Fig. 6.11, p. 57

the earliest possible time the activity can start (*EST*) is week 16, while the latest possible time it can finish (*LFT*) is week 39. Thus, it can be said that:

$$\text{Maximum available time} = 39 - 16 \text{ weeks}$$
$$= 23 \text{ weeks}$$

Now the activity only 'needs' the duration time in order that it can be completed, that is:

$$\text{Necessary time} = 15 \text{ weeks}$$

Thus, the activity can 'expand' or 'move' by $(23 - 15) = 8$ weeks. Any expansion or movement *greater* than this will change the critical path and increase the overall project time. This time of 8 weeks is known as the *total float* possessed by the activity (Fig. 7.2).

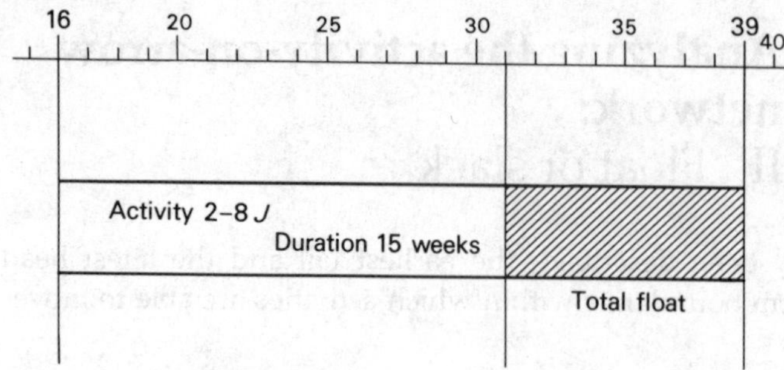

Fig. 7.2

Total float: the total amount an activity may move without affecting the total project time (*TPT*).

It must be realized that the total float is here shown as appearing as time at the *end* of an activity, but that this is not necessarily the case. Float can appear at the beginning of an activity, that is, the starting of the activity can be delayed after the tail node is reached; or it can appear *in* the activity, so that the duration time is increased beyond that initially planned; or it can appear after the activity is finished, while other activities are being concluded to reach the head node.

Free and independent float

Examining activity *K* in the same way it will be seen that:

Maximum available time	= (51 − 38) weeks
	= 13 weeks
Necessary time	= 12 weeks
Hence, float	= 1 week

However, if activity *J* actually absorbs all its float of 8 weeks, event 8 will be reached by week (16 + 15 + 8) = week 39. Thus, activity *K* cannot possibly start until week 39, and

Available time	= (51 − 39) weeks
	= 12 weeks
Necessary time	= 12 weeks
Hence, float	= 0 weeks

so that, if activity *J* absorbs all its float, activity *K* has no float remaining. On the other hand, if activity *J* absorbs only 7 weeks or less of its float, the float in activity *K* remains unaltered at 1 week.

It can thus be said that activity *J* has 8 weeks' *total* float, of which 7 can be used without reducing the float in any succeeding activity. One way of expressing this is to say that there is an interference float of 1 week associated with activity *J*. A more common, and more useful, mode of expression is to say that activity *J* has a *total* float of 8 weeks and a *free* float of 7 weeks.

> *Free float:* the total amount an activity may move without affecting subsequent activities or the *TPT*.

In the planning stage it may be decided to increase the duration time of activity *J* (for example, by reducing the resources allocated to it and thus increasing its performance time). If this is done, then the float available in *previous* activities will be reduced, so that the term *'free'* indicates only that use of the float will not affect any succeeding activities. Cases do arise where the absorption of float affects neither earlier nor later activities, and the float is then said to be *independent*.

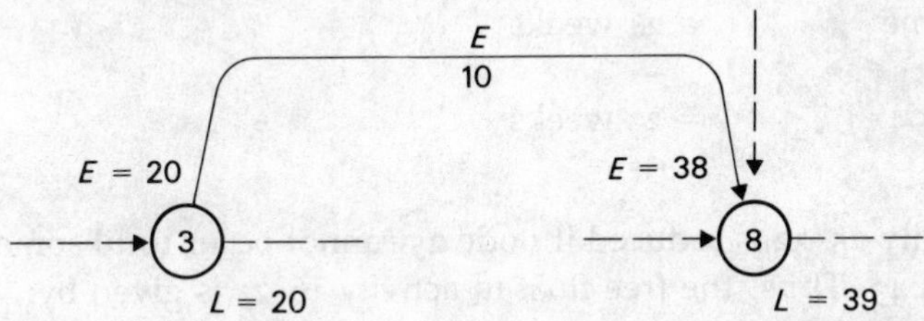

Fig. 7.3 Refer back to Fig. 6.11, p. 57

Activity *E* in Fig. 7.3 has a maximum available time of (39 − 20) weeks = 19 weeks, and a necessary time of 10 weeks, so that the *total* float is 9 weeks. Analyzing as before, it will be found that the *free* float is 8 weeks. If all associated activities take all the float possible, that is, if the tail node is reached as *late* and the head node occurs as *early* as possible, then the time available to *E* is a minimum:

Minimum available time	=	(38 − 20) weeks
	=	18 weeks
Necessary time	=	10 weeks
Hence, *independent* float	=	8 weeks

Independent float: the total amount an activity may move without affecting any other activity either previous or subsequent or increasing the *TPT*.

Summarizing this, we have:

Activity	*Duration*	*Start*		*Finish*		*Float*		
		Early	*Late*	*Early*	*Late*	*Tot.*	*Free*	*Ind.*
3–8	10	20	29	30	39	9	8	8

As a further example of the different types of float, consider Fig. 7.4, which is part of a network not hitherto considered:

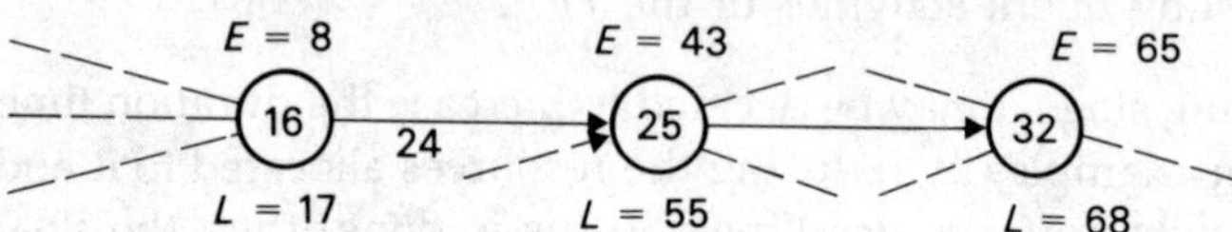

Fig. 7.4

For activity 16–25:

Maximum available time = 47 weeks
Necessary time = 24 weeks

∴ Total float = 23 weeks

Float in activity 25–32 is reduced if node 25 cannot occur until some time *after* week 43. Thus, the free float in activity 16–25 is given by:

Earliest event time (*EET*) for node 16
+ duration time for activity 16–25
+ free float for activity 16–25
= *EET* for node 25,

i.e. $8 + 24 + (\text{free float})_{16-25} = 43$
i.e. $(\text{free float})_{16-25} = 43 - 32$ weeks
$= 11$ weeks.

Minimum available time = 26 weeks
Necessary time = 24 weeks

∴ Independent float = 2 weeks

which, summarized as before, gives:

Activity	*Duration*	*Start*		*Finish*		*Float*		
		Early	*Late*	*Early*	*Late*	*Tot.*	*Free*	*Ind.*
16–25	24	8	31	32	55	23	11	2

Negative float

It is sometimes convenient to compare the overall project time with a target or acceptable time, and this can be very conveniently done by 'turning round' at this target time. Thus, if the $E =$ $L =$ technique is used, the target time is inserted at the final $L =$ figure. The latest event times (*LET*s) are then calculated from this final $L =$ time, and float is again extracted. If the target time is greater than the *TPT*, then *all* activities will have positive float, while if the target time is *less* than the *TPT*, the critical path, and possibly some other activities, will have *negative* float. This negative float is the time by which its associated activity must be reduced for the project to meet the target time. This extended concept of float then gives a precise definition of the critical path:

> The critical path in a network is that path which has least float.

Significance of float

The importance of knowing the types of float depends upon the use made of the information. For example, if it is desired to reduce the effort on a non-critical activity, thus increasing its duration time but releasing effort for use elsewhere, then independent float can be used without replanning any other activities. On the other hand, free float can be used without affecting subsequent activities, while total float may affect both previous and subsequent activities. Negative float indicates the reduction in duration time required to meet a target date. Many workers find that in practice all that is necessary is to calculate and use total float.

Slack

A different expression of the ability of activities to move is given by considering the head and tail events. These have 'earliest' and 'latest' times, and slack is the difference between these times. Thus, for event 2 (Fig. 7.5):

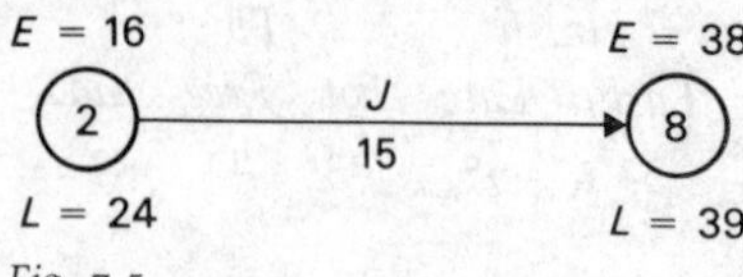

Fig. 7.5

where the *EET* is 16 weeks and the *LET* is 24 weeks, the slack is said to be 8 weeks, and for event 8 it is 1 week. Remembering that the beginning of an activity is the tail and the end is the head, it can be said that activity 2–8 has a tail slack of 8 weeks and a head slack of 1 week. Using the tail slack of an activity affects in general the slack in both earlier and later events, while using the head slack affects in general only the subsequent events.

The relationship between float and slack is:

Free float = Total float − head slack
Independent float = Free float − tail slack

Applying this to activity 2–8 we have:

Total float = 39 − 16 − 15 weeks = 8 weeks
Head slack = 1 week
Tail slack = 8 weeks
∴ Free float = Total float − head slack
= (8 − 1) weeks
= 7 weeks

Independent float = free float − tail slack = 7 − 8 weeks = −1 week, which for practical purposes is taken to be zero.

As with float, slack can be a measure of the acceptability of the project as planned. Thus, if the critical path length is 51 weeks and the maximum acceptable time is 41 weeks, then events on the critical path are said to have −10 weeks' slack, or 10 weeks' negative slack. This usage is quite convenient when a project is actually running and slack can be calculated from actual rather than predicted duration times. Should slack be positive then it is possible to meet the accepted overall time without replanning. If slack is negative replanning is essential to return the project to its previously agreed overall time.

Note: Do not confuse float and slack. Slack refers to *events*, float refers to *activities*. Some (not all) US books use the terms as synonyms, while some reverse the usage prescribed here. The present terminology is that which is recommended by BS 4335:1972. In all the situations in which the author has worked 'slack' has not been found to be a useful expression of available time.

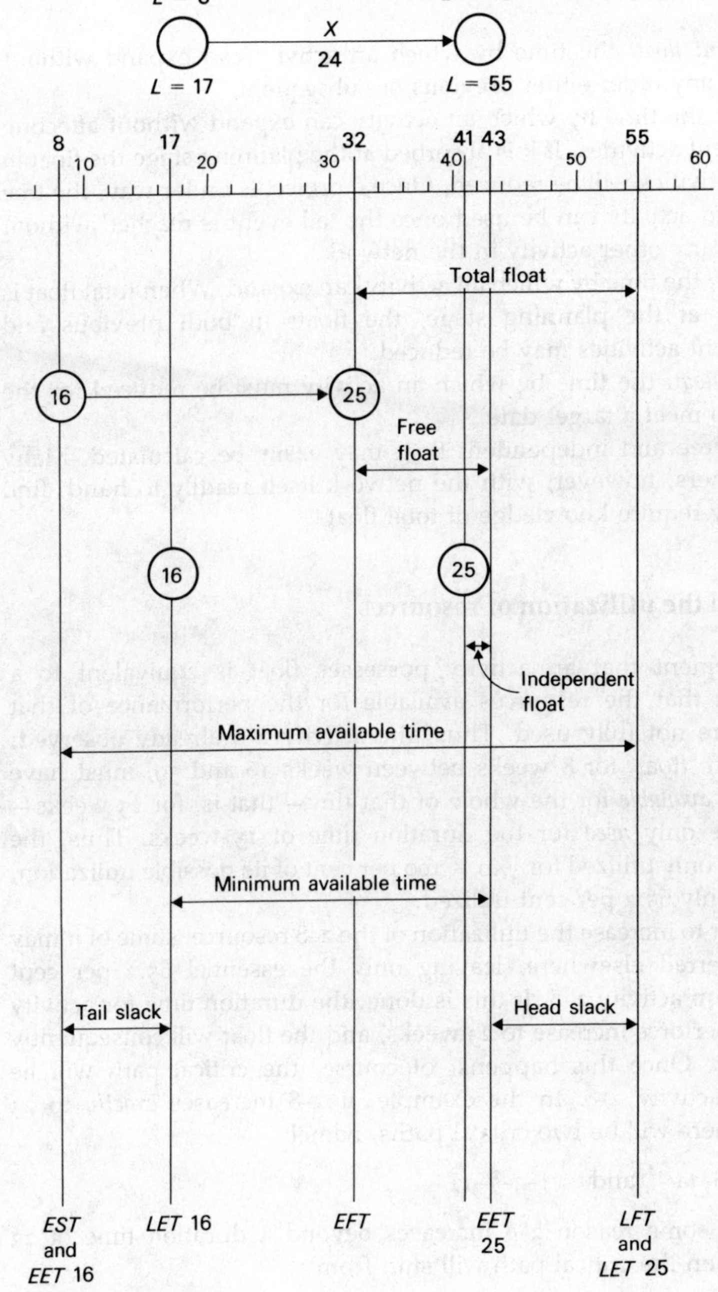

Fig. 7.6 A graphical summary

Float: a summary

Independent float: the time by which an activity can expand without affecting any other either previous or subsequent.
Free float: the time by which an activity can expand without affecting subsequent activities. If it is absorbed at the planning stage the float in earlier activities will be reduced. Once a project is under way, the free float in an activity can be used once the tail event is reached without affecting any other activity in the network.
Total float: the time by which an activity can expand. When total float is absorbed at the planning stage, the floats in both previous and subsequent activities may be reduced.
Negative float: the time by which an activity must be reduced for the project to meet a target date.

Note: Free and independent float may easily be calculated. Many practitioners, however, with the network itself readily to hand, find they only require knowledge of total float.

Float and the utilization of resources

The statement that an activity possesses float is equivalent to a statement that the resources available for the performance of that activity are not fully used. Thus, the activity 2–8 already observed, which can 'float' for 8 weeks between weeks 16 and 39, must have resources *available* for the whole of that time—that is, for 23 weeks—which are only *used* for the duration time of 15 weeks. Thus, the activity is only utilized for $15/23 \times 100$ per cent of its possible utilization, i.e. it is only 65.2 per cent utilized.

In order to increase the utilization of the 2–8 resource, some of it may be transferred elsewhere, leaving only the essential 65.2 per cent resource on activity 2–8. If this is done, the duration time for activity 2–8 will perforce increase to 23 weeks, and the float will consequently disappear. Once this happens, of course, the critical path will lie through activity 2–8. In the example, if 2–8 increases *exactly* to 23 weeks, there will be two critical paths, namely:

1–2–8–11 and 1–3–7–11

or, if for some reason 2–8 increases beyond a duration time of 23 weeks, then the critical path will shift from

1–3–7–11 to 1–2–8–11.

If *all* resources are fully utilized, the whole network becomes critical. At first sight this might appear to be a highly desirable situation in that there are no idle resources. It must be remembered, however, that idle resources represent some degree of flexibility in a project, and that to remove this flexibility might result in a state of crisis that could have been avoided—or at least alleviated—if some float had been available.

Thus, we have two general comments:

(1) Float represents under-utilized resources.
(2) Float represents flexibility.

Rules for calculating float

Total float: subtract the earliest time for the preceding node from the latest time for the succeeding node, and from this difference subtract the duration time. Example: for activity 3–8 (p. 57):

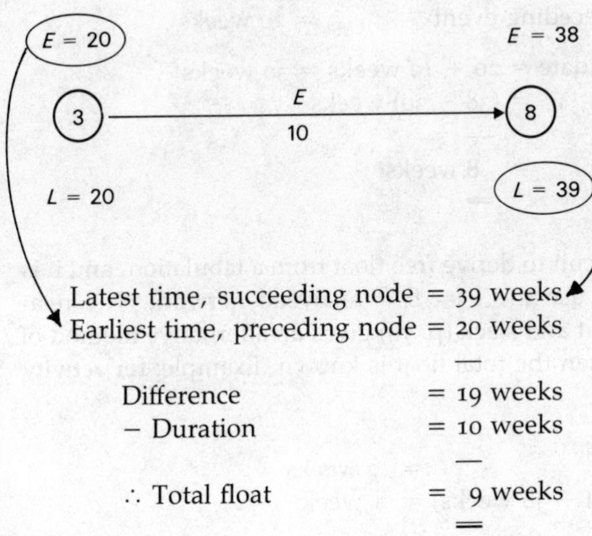

Note: When setting out an analysis in tabular form a useful alternative statement of the above rule is given by:

Total float: 'Latest start date of activity minus earliest start date of activity. (May be negative.)'

Thus, for activity 3–8:

Latest start date	= (39 − 10) weeks =	29 weeks
Earliest start date	=	20 weeks
∴ Total float	=	9 weeks

Free float: free float can be calculated by: 'Earliest date of succeeding event minus earlier finish date of activity.' Example: for activity 3–8:

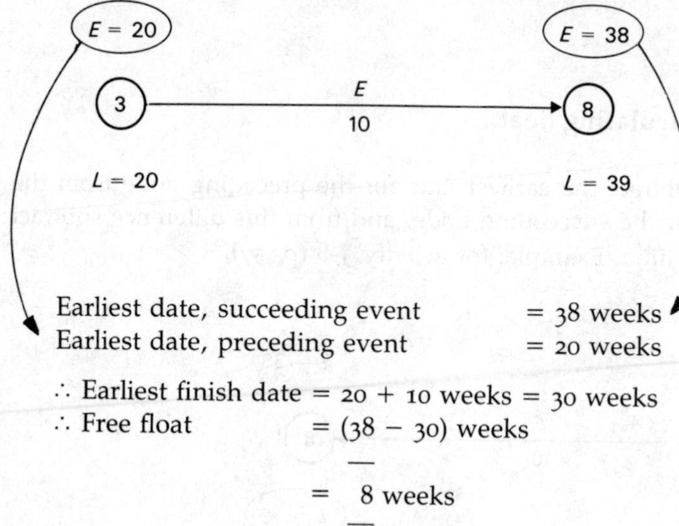

Earliest date, succeeding event	=	38 weeks
Earliest date, preceding event	=	20 weeks
∴ Earliest finish date = 20 + 10 weeks	=	30 weeks
∴ Free float	=	(38 − 30) weeks
	=	8 weeks

Note: It is very difficult to derive free float from a tabulation, and it is always necessary to use the *E* = *L* = analyzed network. The relationship between float and slack (p. 70) gives an alternative method of calculation useful when the total float is known. Example: for activity 3–8:

Total float	=	9 weeks
− Head slack (39 − 38 weeks)	=	1 week
∴ Free float	=	8 weeks

Independent float: this is derived from: 'Independent float is earliest date of succeeding node minus latest date of preceding node minus activity duration.' (If negative, the independent float is taken as zero.) Example: for activity 3–8:

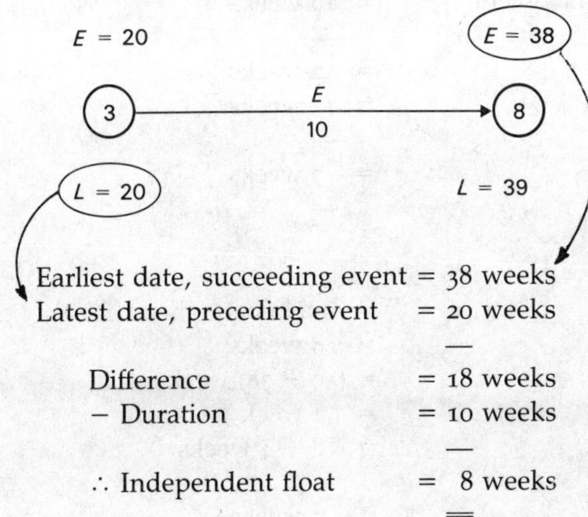

Earliest date, succeeding event	=	38 weeks
Latest date, preceding event	=	20 weeks
Difference	=	18 weeks
− Duration	=	10 weeks
∴ Independent float	=	8 weeks

Note: As with free float, calculation of independent float requires reference to the $E = L =$ analyzed network. Again, the relationship between float and slack (p. 70) gives an alternative method of calculation that is useful when free float is known. Example: for activity 3–8:

Free float	=	8 weeks
− Tail slack = (20 − 20 weeks)	=	0 weeks
∴ Independent float	=	8 weeks

Further illustration of float calculations

Activity 2–8, *J*:

Total float:

Latest start time	=	24 weeks
Earliest start time	=	16 weeks
∴ Total float	=	8 weeks

Free float:

Earliest time head event	=	38 weeks
Earliest time tail event	=	16 weeks
Difference	=	22 weeks
− Duration	=	15 weeks
∴ Free float	=	7 weeks

Alternatively:

Free float	=	total float − head slack
Total float	=	8 weeks
Head slack	=	(39 − 38) weeks
	=	1 week
∴ Free float	=	8 − 1 weeks
	=	7 weeks

Independent float:

Earliest time, head event	=	38 weeks
Latest time, tail event	=	24 weeks
Difference	=	14 weeks
− Duration	=	15 weeks
∴ Independent float	=	−1 week

which for practical purposes is written:

Independent float	=	0 weeks

Alternatively:

Independent float	=	free float − tail slack
Free float	=	7 weeks
− Tail slack = (24 − 16 weeks)	=	8 weeks
∴ Independent float	=	−1 week

which is written as:

Independent float = 0 weeks.

Activity 8–11*K*:

Total float = (51 − 38) − 12 weeks = 1 week
Free float = Total float − head slack = 1 − 0 = 1 week
Independent float = Free float − tail slack = 1 − 1 = 0

Generalized rules for analysis

Expressing the rules generally in mathematical terms we have, for the generalized activity *i–j* of duration *D* (Fig. 7.7) where the tail event of the activity is *i*, and the head event of the activity is *j*, and the earliest and latest event times are denoted by subscripts *E* and *L*:

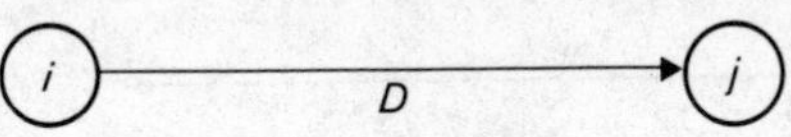

Fig. 7.7

Earliest time of tail event $i = i_E$
Latest time of tail event $i = i_L$
Earliest time of head event $j = j_E$
Latest time of head event $j = j_L$

Then for activity *i–j*:

Earliest start time *EST* $= i_E$
Latest start time *LST* $= j_L - D$
Earliest finish time *EFT* $= i_E + D$
Latest finish time *LFT* $= j_L$
Total float $= j_L - i_E - D$
Free float $= j_E - i_E - D$
Independent float $= j_E - i_L - D$
Event slack for event $= i_L - i_E$

or, having calculated total float (*TF*):

free float = total float − head slack
(*FF*= *TF* − *HS*)
independent float= free float − tail slack
(*IF*= *FF* − *TS*)

Using the above rules for the sample network already discussed, we have the following (Fig. 7.8):

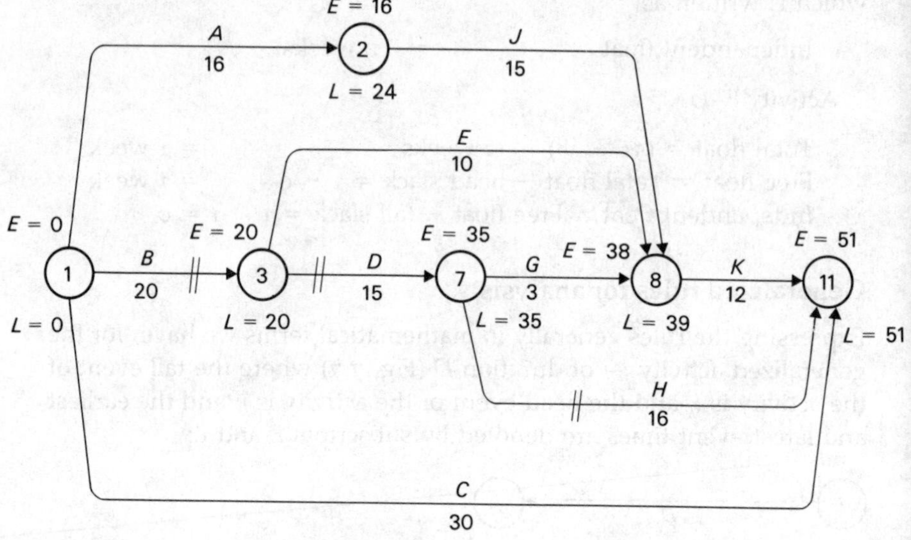

Fig. 7.8

No.	Activity Description	Duration (weeks)	Start Early	Start Late	Finish Early	Finish Late	Float Tot.	Float Free	Float Ind.
1– 2	A	16	0	8	16	24	8	0	0
1– 3	B	20	0	0	20	20	0	0	0
1–11	C	30	0	21	30	51	21	21	21
2– 8	J	15	16	24	31	39	8	7	0
3– 7	D	15	20	20	35	35	0	0	0
3– 8	E	10	20	29	30	39	9	8	8
7– 8	G	3	35	36	38	39	1	0	0
7–11	H	16	35	35	51	51	0	0	0
8–11	K	12	38	39	50	51	1	1	0

Completing an analysis table

In practice, it will be found most convenient, when compiling an analysis such as the above to proceed as follows:

(1) Fill in activity numbers, ensuring that every arrow in the network is represented by a head and tail number.
(2) Fill in duration times.

(3) Write down *EST* from the analyzed network, using the *E* of the tail event.
(4) Write down *LFT* also from the network, using the *L* of the head event.
(5) Calculate *LST* from above by subtracting duration time from *LFT*.
(6) Calculate *EFT* from above by adding duration time to *EST*.
(7) Calculate total float by subtracting *EFT* from *LFT*.
(8) Calculate free and independent float from network using rules on pp. 73–75 or relationship on p. 77.

Verification of an analysis

When an analysis has been made there are a number of simple checks that can be carried out which, though not in themselves conclusive, can bring some arithmetical and recording errors to light. For example:

(1) All activities with the same tail number have the same *EST*.
(2) All activities with the same head number have the same *LFT*.
(3) *EST*s are never larger than *LST*s.
(4) *EFT*s are never larger than *LFT*s.
(5) Start times are always earlier than corresponding finish times.
(6) Free float can never exceed total float.
(7) Independent float can never exceed free float.
(8) Total float is the difference between *EST* and *LST*.

Intermediate imposed times

Circumstances sometimes require that events other than, or as well as, the final event should take place at particular times. Should this be so, then these 'scheduled' or 'imposed' times can be inserted into the network by means of a solid arrow-head, surmounted by the scheduled time. For example, Fig. 7.9

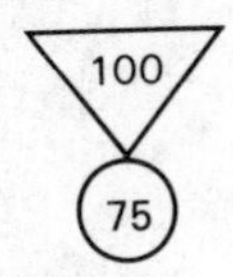

Budget approved

Fig. 7.9

indicates that event 75 'budget approved' must take place by time 100, that is, that all activities with a head number 75 must be complete by time 100. An arrow-head inverted below the event number indicates that activities with tail number 75 cannot start until time 100.

The scheduled time having been inserted, analyses can now be carried out (*i*) between the beginning and end of the network and (*ii*) between the beginning and/or end and the intermediate scheduled points, treating these points as if they were starts or finishes. This can then give rise to critical paths other than the main critical path, and these are known as secondary, tertiary and so on, each with its own set of floats. Once such an analysis has been carried out, the meanings of the various factors will become immediately obvious.

8 Ladder activities

A method of representing overlapping activities, sponsored and used by ICL is the so-called '*ladder* method'. In this, *restraint* arrows are used to indicate the minimum intervals that must elapse between the start of one activity and the start of the next, and the finish of one activity and the finish of the next. For example, if *P* has a duration time of 3, and *S* may start when 1 unit of time of *P* is complete, and at least 2 units of time of *S* must elapse after the completion of *P*, *S* having a total duration of 5, then the ladder representation would be as shown in Fig. 8.1.

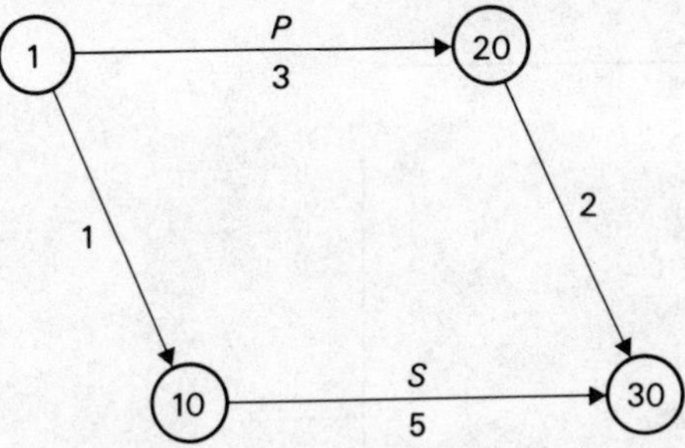

Fig. 8.1

The arrows 1–10 and 20–30 represent the restraints imposed upon *S* and *P*. These constraints are sometimes known as 'lead' and 'lag' activities and are labelled accordingly, but it is helpful to indicate their physical meaning by labelling them 'START' and 'FINISH' (Fig. 8.2).

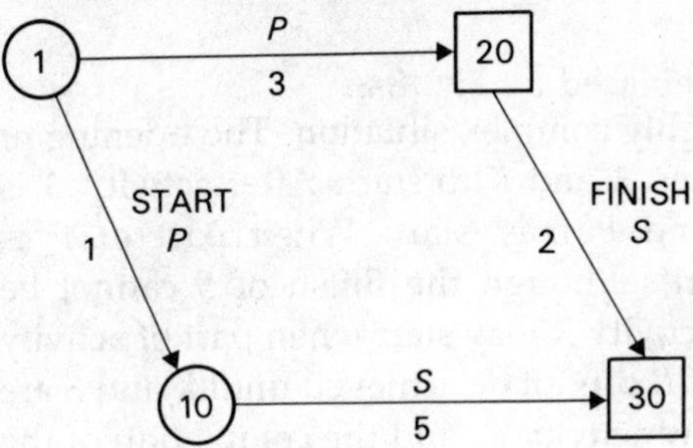

Fig. 8.2

As the computational methods for ladders differ slightly from a conventional CPA, it is as well to use a different node symbol as a 'warning flag'. Ladders of this sort can be built up as the situation demands.

The different computational needs also require an 'isolation' of the ladder from the rest of the network. This isolation generates two special 'ladder' rules:

(1) A ladder activity, and its associated 'start' activity must have a common tail node, and no other activity may emerge from the node.
(2) A ladder activity, and its associated 'finish' activity must have a common head node, and no other activity may enter that node.

Thus, Fig. 8.3

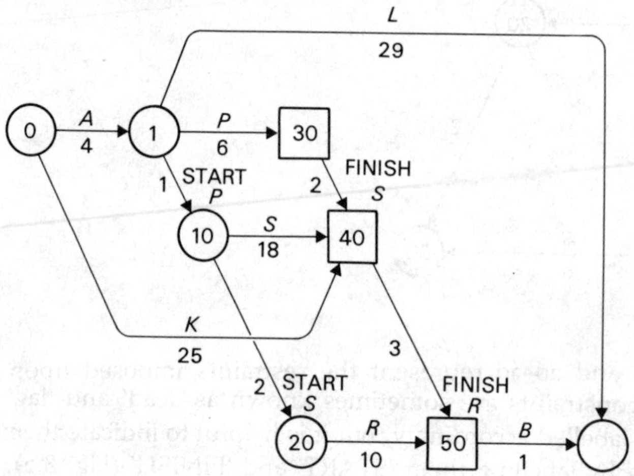

Fig. 8.3

is unacceptable, and must be replaced by Fig. 8.4.

Note: Fig. 8.4 represents a highly complex situation. The opening of the project permits two activities *A* and *K* to start. After activity *A* is completed, two activities *L* and *P* may start. When part of *P* is completed, activity *S* may start, although the finish of *S* cannot be achieved until *P* is complete. Activity *R* may start when part of activity *S* is completed, but the finish of *R* cannot be achieved until *K* and *S* are complete. Activity *B* must follow activity *R*, and the completion of the project requires the completion of both *B* and *L*.

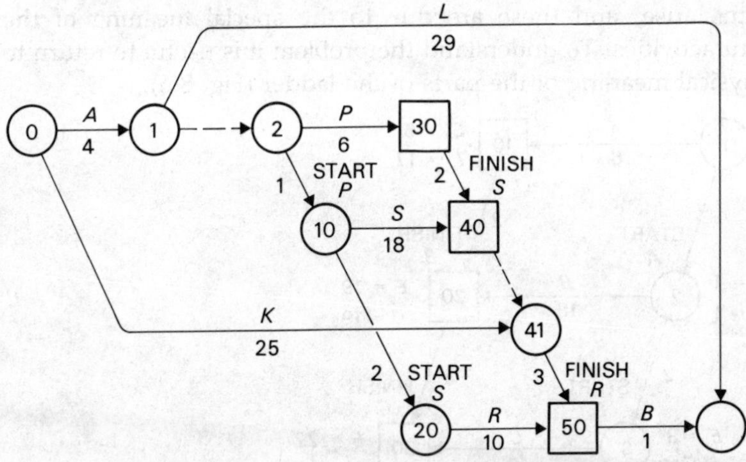

Fig. 8.4

Represented by conventional means, the network would appear as in Fig. 8.5.

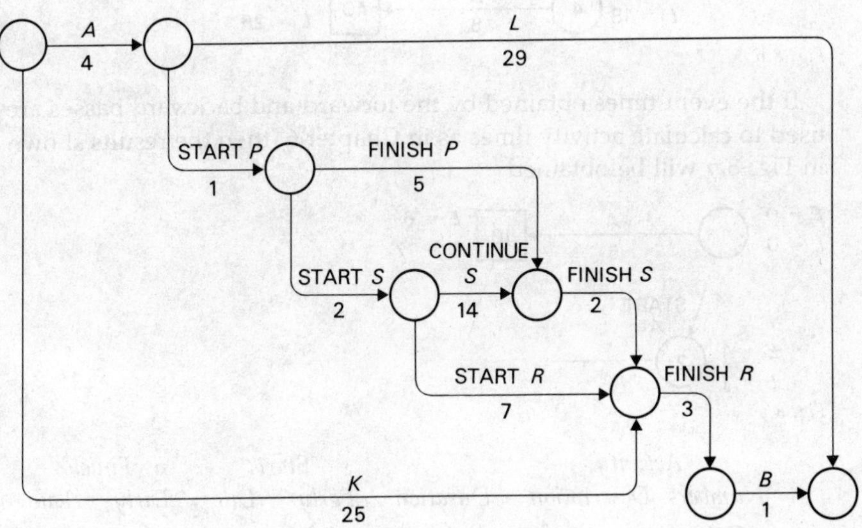

Fig. 8.5

Calculation of ladders

Calculation of a ladder drawn to the above conventions starts with a forward and backward pass carried out in the manner discussed in Chapter 6. It is in the subsequent calculation of activity times that problems arise, and these are due to the special meaning of the restraint activities. To understand the problem it is useful to return to the physical meaning of the parts of the ladder (Fig. 8.6).

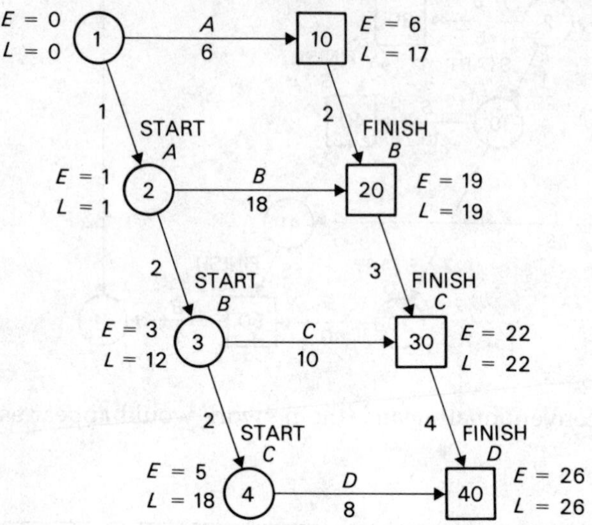

Fig. 8.6

If the event times obtained by the forward and backward passes are used to calculate activity times as in Chapter 6, then the results shown in Fig. 8.7 will be obtained:

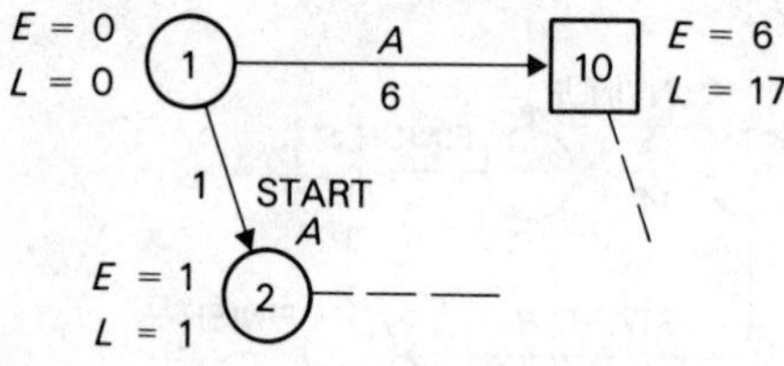

Fig. 8.7

Activity			*Start*		*Finish*	
Numbers	*Description*	*Duration*	*Early*	*Late*	*Early*	*Late*
1–10	*A*	6	0	11	6	17
1– 2	START *A*	1	0	0	1	1

Apparently 'START *A*' will finish at time 1, while *A* could start at time 11, that is, the whole activity could start after a part of itself had finished, something that is difficult to envisage.

To understand the physical nature of the situation, consider the Gantt chart for the network in the 'early' position (Fig. 8.8). Activities

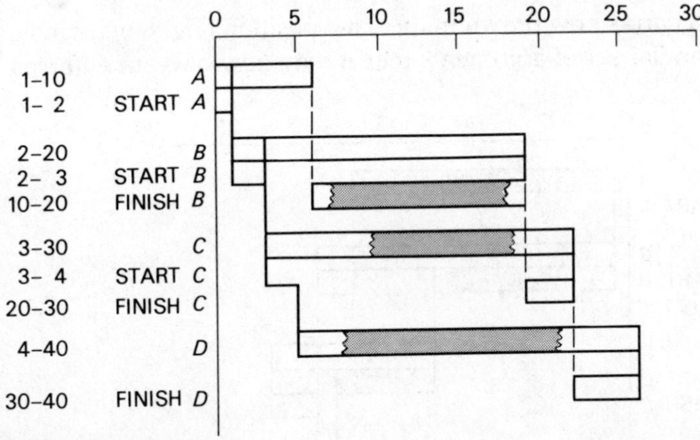

Fig. 8.8 Early position

1–10 and 1–2 must have a common start since they are inseparably joined, as are activities 2–20 and 2–3, and 3–30 and 3–4. Similarly, the finishes of activities 2–20 and 10–20 must be the same, as must be 3–30 and 20–30, and 4–40 and 30–40. If activity 10–20 is now considered, it will be seen that it can *start* after *A* has finished, that is, at time 6, but that it *must* finish at the same time as *B*, that is, at time 19. The duration time of activity 10–20 is only 2 units of time, so that there must be some 'dead' time of magnitude (19–6–2) = 11 units of time incorporated into activity 10–20. This waste time (or 'enforced idle time') is shown by the 'dotted' shading and the ragged ends to the activity 10–20 bar.

In the same way, C must incorporate some enforced idle time (*EIT*) in the early position, this being of magnitude (22–3–10) = 9 units of time. *D* will also have *EIT* of (26–5–8) = 13 units of time. The above arguments may be summarized in the following table:

Numbers	*Activity Description*	*Duration*	*Start*	*Finish*	*EIT*
1–10	*A*	6	0	6	0
1– 2	START *A*	1	0	1	0
2–20	*B*	18	1	19	0
2– 3	START *B*	2	1	3	0
10–20	FINISH *B*	2	6	19	11
3–30	C	10	3	22	9
2– 4	START C	2	3	5	0
20–30	FINISH C	3	19	22	0
4–40	*D*	8	5	26	13
30–40	FINISH *D*	4	22	26	0

If the Gantt chart is now drawn in the 'late' position (Fig. 8.9) it can be seen, by a similar set of arguments that *A* may now have an enforced

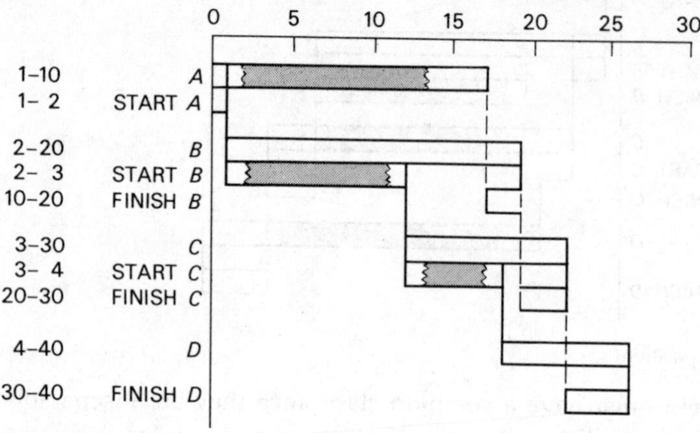

Fig. 8.9 Late position

idle time of 11 units of time, its start being located by the start of the whole project, while its finish may 'drift' to a position located by 'FINISH *B*'. 'START *B*' may have an *EIT* of 9 units of time and 'START *C*' an *EIT* of 4 units of time. The 'late' position can be summarized as follows:

Activity Numbers	*Activity* Description	*Duration*	*Start*	*Finish*	*EIT*
1–10	*A*	6	0	17	11
1– 2	START *A*	1	0	1	0
2–20	*B*	18	1	19	0
2– 3	START *B*	2	1	12	9
10–20	FINISH *B*	2	17	19	0
3–30	*C*	10	12	22	0
3– 4	START *C*	2	12	18	4
20–30	FINISH *C*	3	19	22	0
4–40	*D*	8	18	26	0
30–40	FINISH *D*	4	22	26	0

Examination of the two tables in conjunction with the forward and backward pass will reveal that the *EST* and *EFT* for any activity are given by the *E* values at the tail and the head of the activity, and that the *LST* and *LFT* are the corresponding *L* values. In both cases the *EIT* is given by subtracting the duration time of the activity from the difference between the start and finish times. Amalgamating the two tables into one table, the following is obtained:

Activity			*Early*			*Late*		
Numbers	*Description*	*Duration*	*Start*	*Finish*	*EIT*	*Start*	*Finish*	*EIT*
1–10	*A*	6	0	6	0	0	17	11
1– 2	START *A*	1	0	1	0	0	1	0
2–20	*B*	18	1	19	0	1	19	0
2– 3	START *B*	2	1	3	0	1	12	9
10–20	FINISH *B*	2	6	19	11	17	19	0
3–30	*C*	10	3	22	9	12	22	0
3– 4	START *C*	2	3	5	0	12	18	4
20–30	FINISH *C*	3	19	22	0	19	22	0
4–40	*D*	8	5	26	13	18	26	0
30–40	FINISH *D*	4	22	26	0	22	26	0

The critical path is the path that has minimum float, or in this case, the special float defined as enforced idle time, and both the 'early' and the 'late' situations need to be examined. In this case the critical path is 1–2–20–30–40. Note that in the special case of ladders, using the isolating conventions above, the critical path is defined as that path which lies between events where earliest and latest times are equal.

The ladder in practice

The above discussion is unreal in the sense that it is unlikely that a task will be planned to have an enforced idle time. In practice, enforced idle time having been identified, the resources on the activities concerned will, if possible be adjusted so that the duration time of the activity will increase to absorb the idle time. Thus, if the 'earliest' situation is acceptable, then attempts are likely to be made to increase, for example, *C* from a duration time of 10 to a duration time of 19 by diluting the resources applied. This will have the effect of causing *C* to become critical and, as usual, the more efficient the deployment of resources, the less flexibility is available.

Note: The ladder convention is not easy to use, particularly in the absence of resource allocation and modification facilities. Used with care, however, it can provide a useful short-hand method of dealing with a difficult situation as in the diagram on p. 83. When calculating by computer it is, of course, necessary to code the components of the ladder in accordance with the program user's instructions.

9 Analyzing the activity-on-node network: I Isolating the critical path

The network used throughout this text is in MoP (method of potentials) terms, as shown in Fig. 9.1, and all calculations will be carried out on this network.

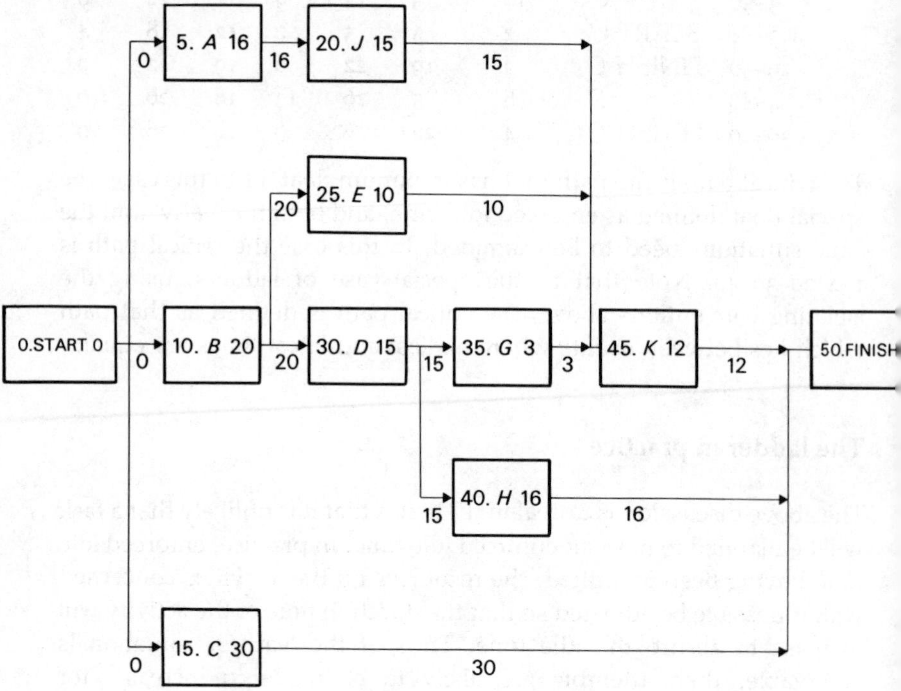

Fig. 9.1 Time in weeks

Calculating the total project time (*TPT*)

The *TPT* is the shortest time in which the project can be completed, and this is determined by the sequence (or sequences) of activities known as the critical path(s). To calculate the *TPT* a *forward pass* is carried out whereby the earliest start time (*EST*) of each activity is calculated. This *EST* is generally written in or on the top left-hand corner of the node.

The forward pass

(1) Start at the beginning of the network, that is at the START activity.

(2) Assign to the START node an *EST* of 0. This assumes that the whole project can start *now*. *Note*: if there is a known interval of time before which the project can start—say *X* weeks—then an *EST* of *X* *can* be assigned to the start node. It is usually simpler, however, to assign an *EST* of zero to the START activity and then add *X* to all times subsequently calculated.

(3) Proceed to each activity in turn and calculate from the *EST* of the preceding activity and the dependency time its *EST* as shown in Fig. 9.2.

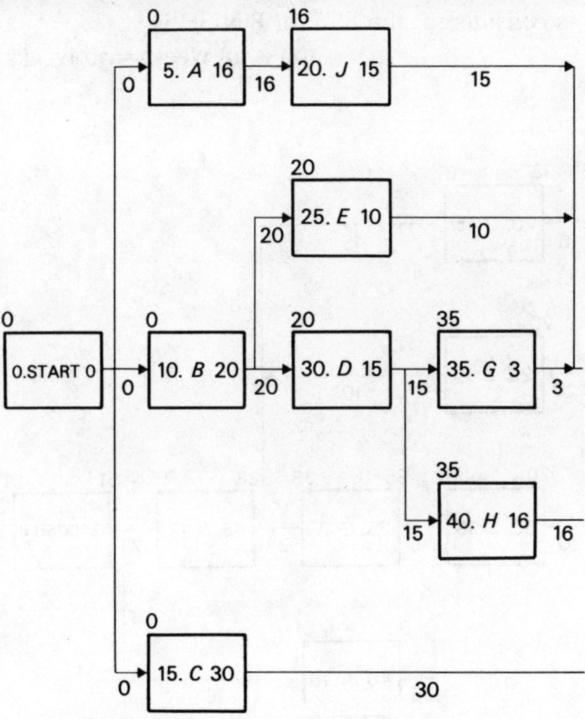

Fig. 9.2 Partly calculated network

(4) Three activities, *J*, *E* and *G*, converge upon activity *K*. The *EST* of activity *K* is therefore given by the largest of the sum of the *EST* and the dependency time for each immediately preceding activity. Thus:

If *only* J is considered, the *EST* for K is

(16 + 15) weeks = 31 weeks

If *only* E is considered, the *EST* for K is

(20 + 10) weeks = 30 weeks

If *only* G is considered, the *EST* for K is

(35 + 3) weeks = 38 weeks

Since J, E and G must *all* be considered the *EST* for K is 38 weeks.

Similarly to FINISH activity:

If *only* K is considered, the *EST* for FINISH is

(38 + 12) weeks = 50 weeks

If *only* H is considered, the *EST* for FINISH is

(35 + 16) weeks = 51 weeks

If *only* C is considered, the *EST* for FINISH is

(0 + 30) weeks = 30 weeks

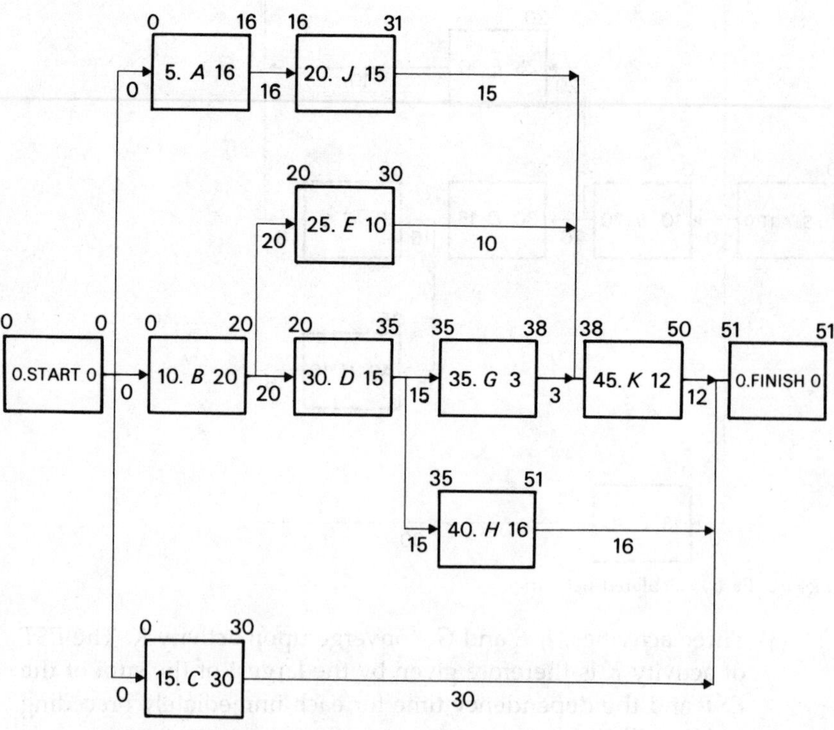

Fig. 9.3 Network with earliest times calculated

Since *K*, *H* and *C* must all be considered the *EST* for FINISH is 51 weeks. As the FINISH activity has zero duration, an *EST* for the FINISH activity of 51 weeks means that the earliest time in which the project can be completed—the *TPT*—is 51 weeks.

(5) From the *EST* and the duration time, calculate the earliest finish time (*EFT*).

Earliest finish time = Earliest start time + duration

$$EFT = EST + d$$

Thus, for activity *K*:

$$EFT = (38 + 12) \text{ weeks} = 50 \text{ weeks}$$

The *EFT* is generally placed in or on the top right-hand corner, thus having the earliest times on the top line as shown in Fig. 9.3.

(6) It will be seen that the *EFT* of the FINISH node is 51, which is, therefore, the *TPT*.

Isolating the critical path

The backward pass

The critical path can be isolated by carrying out a *backward pass* whereby the latest start time (*LST*) of each activity is derived. This is generally placed in or under the bottom left-hand corner of the node.

Note: This differs from the backward pass calculation in arrow-on-arrow (AoA) where the latest *finish* time (*LFT*) for each activity is calculated. This difference arises from the fact that in MoP the dependency arrow sets the difference between the start of an activity and the start of immediately dependent activities.

(7) Start now at the end of the network, that is, at the FINISH activity.

(8) Assign to this activity an *LST* equal to its *EST*. This is equivalent to a statement that the project will be completed as quickly as possible.

(9) By successively subtracting dependency times from *LST*s calculate the *LST* of each activity. Thus, for activity *K*:

$$LST(K) = LST\ (\text{FINISH}) - 12 = (51 - 12) \text{ weeks} = 39 \text{ weeks}$$

and for activity:

$$LST(G) = LST(K) - 3 \qquad = (39 - 3) \text{ weeks} = 36 \text{ weeks}$$

This gives the results as shown in Fig. 9.4.

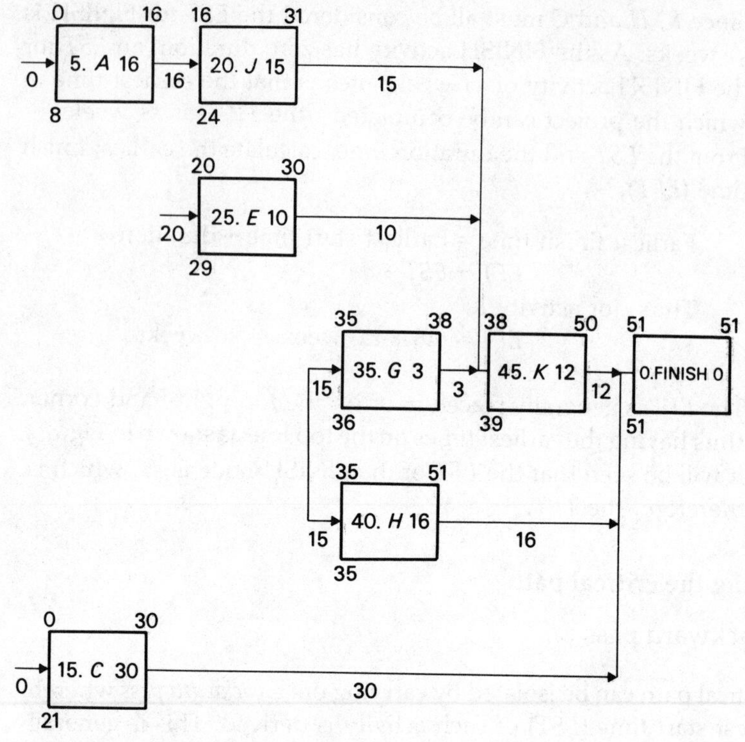

Fig. 9.4 Partly calculated backward pass

(10) Two activities, *G* and *H* emerge from activity *D*. The earliest finish time (*EFT*) of activity *D* is therefore given by the smaller of the differences between the *LST* and the duration times. Thus:

If *only* G is considered, the *LST* for *D* is

(36 − 15) weeks = 21 weeks

If *only H* is considered, the *LST* for *D* is

(35 − 15) weeks = 20 weeks

Since *both* G and *H* must be considered, the *LST* for *D* is 20 weeks.

Similarly for activity *B*:

If *only E* is considered, the *LST* for *B* is

(29 − 20) weeks = 9 weeks

If *only D* is considered, the *LST* for *B* is

(20 − 20) weeks = 0 weeks

Since *both* *E* and *D* must be considered, the *LST* for *D* is 0 weeks.

Finally, for START activity:

If *only A* is considered, the *LST* for START =
(9 − 0) weeks = 9 weeks

If *only B* is considered, the *LST* for START =
(20 − 20) weeks = 0 weeks

If *only C* is considered, the *LST* for START =
(21 − 0) weeks = 21 weeks

Since *AB* and *C* must all be considered, the *LST* for START = 0.

Note: (i) The 'turn-round' time at FINISH was equal to the *TPT*—hence the result of the backward pass must be the same as the *EST* of the START activity.

(ii) The forward and backward passes can be summarized:

Forward pass—add and choose the latest figure.
Backward pass—subtract and choose the earliest figure.

The *LST* is generally placed in or under the bottom left-hand corner of the node.

(11) Given the *LST* of an activity and its duration, the *LFT* can be calculated:

Latest finishing time = Latest starting time + duration

$$LFT = LST + d$$

Thus, for activity *K*:

$$LFT = (39 + 12) \text{ weeks} = 51 \text{ weeks}$$

The *LFT* is generally placed in or under the bottom right-hand corner of the node, thus having the latest times at the bottom of the node. The complete calculated network is shown in Fig. 9.5.

Note: The location of the various start and finish times can be remembered:

Start times at the start of the node.
Finish times at the finish of the node.
Earliest times above latest times.

(12) The critical path lies along those activities where there is no 'spare time' and, therefore, in which the *EST and LST* (or *EFT* and *LFT*) are the same. This test only applies if the *EST* and *LFT*

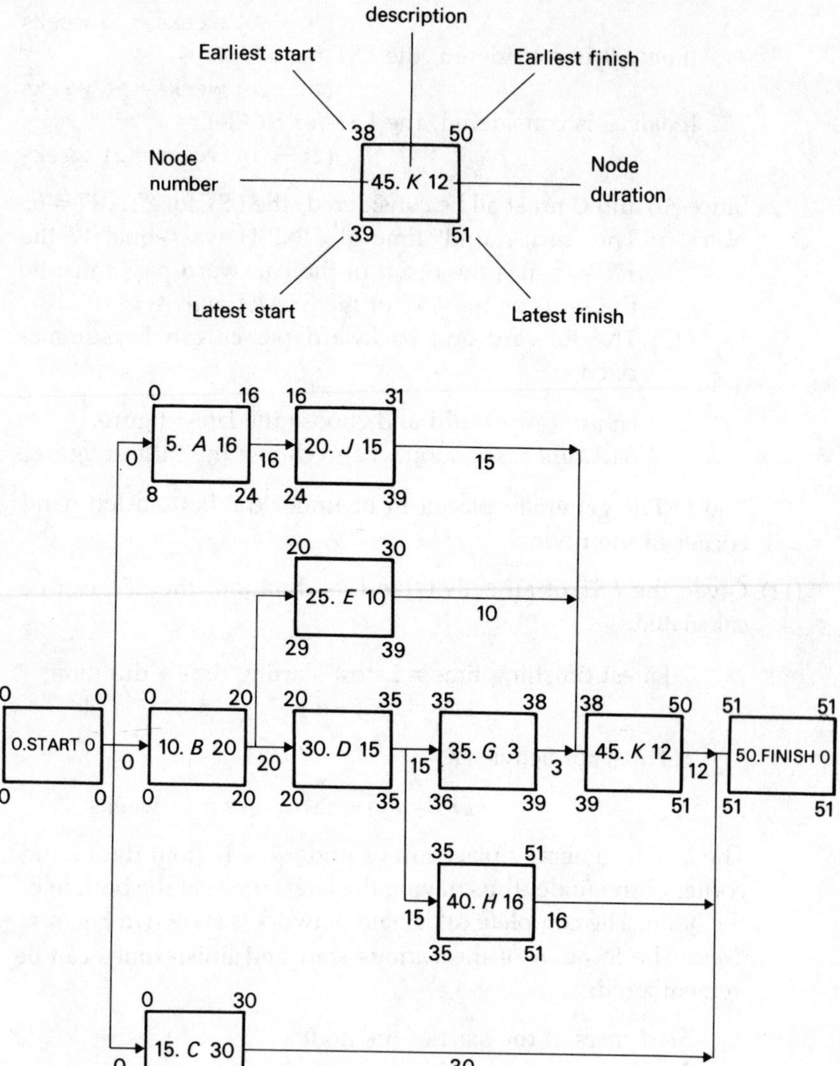

Fig. 9.5 Fully calculated network

at the finish nodes are identical—the most usual situation. In the specimen network, therefore, the critical path is:

Start—activity *B*—activity *D*—activity *H*—Finish

The various times for the network can be tabulated.

Activity					
	Duration	*Start times*		*Finish times*	
Description	*(weeks)*	*Earliest*	*Latest*	*Earliest*	*Latest*
A	16	0	8	16	24
B	20	0	0	20	20
C	30	0	21	30	51
D	15	20	20	35	35
E	10	20	29	30	39
G	3	35	36	38	39
H	16	35	35	51	51
J	15	16	24	31	39
K	12	38	39	50	51

The AoN node in practice

The AoN node is undoubtedly a very convenient and tidy way of assembling all the information concerning an activity *on the network itself*, something that is virtually impossible in AoA. However, the node itself is tedious to draw. The author, being both a poor draughtsman and always anxious to avoid work, has found that pre-printed nodes upon self-adhesive labels simplify the drawing of the network immensely. These labels are easily and cheaply purchased from any of the many jobbing printers who undertake the printing of self-adhesive labels.

In practice, the author fills in the activity description on the label, and disposes the various labels upon a large sheet of paper. The activities are then moved until the logic is correctly represented, and a tidying up, using a straight edge, achieves a pleasing display. The back of each label is then peeled off and the node located firmly in position. All that is then required is the joining up of the various nodes by dependency arrows—a relatively easy task. The author has not yet found a source of transparent labels that can be written upon, so that the drawing must either be re-drawn by a junior upon dye-line paper, or some form of xerography used.

Other workers record the use of large sheets of paper with a faint grid of nodes printed upon it. Activities are located appropriately, the nodes lined in and dependency arrows drawn in the spaces left between the nodes.

Matrix method of expressing and analyzing MoP diagrams

Roy has set down a tabular method of expressing and analyzing an MoP diagram which does not require that the diagram itself should be drawn as a collection of arcs and nodes. A matrix can be used for the same purpose, and this has considerable advantages over the tabular form in clarity and simplicity of construction. To illustrate the method, the same example as previously discussed will be used, although it must be emphasized that the network need not be drawn as on p. 88, the matrix itself can be used to supply all the logical and time dependencies.

Step 1

List all activities, along with their duration times. Add to this list two other activities, START and FINISH, each of zero duration times:

A–16	*D*–15	*G*–3	START–0
B–20	*E*–10	*H*–16	FINISH–0
C–30	*J*–15	*K*–12	

Step 2

Prepare a square matrix with column headings and row descriptions in accordance with the complete list of step 1. It is convenient to put START and FINISH at the beginning and end of the rows and columns, but otherwise the order in which the activities are written down are of no consequence. The column headings represent SUCCESSOR activities and the row descriptions PREDECESSOR activities (Fig. 9.6).

Step 3

Represent logic by 'boxing' the cell at the intersection of dependent activities. Each column except START and each row except FINISH must contain at least one 'boxed' cell (Fig. 9.7).

PREDECESSOR \ SUCCESSOR		START 0	A 16	E 10	G 3	H 16	B 20	K 12	D 15	J 15	C 30	FINISH 0
START	0											
B	20											
C	30											
H	16											
A	16											
D	15											
J	15											
G	3											
E	10											
K	12											
FINISH	0											

Fig. 9.6

PREDECESSOR \ SUCCESSOR		START 0	A 16	E 10	G 3	H 16	B 20	K 12	D 15	J 15	C 30	FINISH 0
START	0		□				□				□	
B	20			□					□			
C	30											□
H	16											□
A	16									□		
D	15				□	□						
J	15							□				
G	3							□				
E	10							□				
K	12											□
FINISH	0											

Fig. 9.7

Step 4

The logic having been set down, dependency times are added by inserting into the 'boxes' the time that must elapse between the starts of the predecessor activity and the successor activity (Fig. 9.8). Constraints (either positive or negative) are added just as in conventional MoP. In the example being discussed, all dependency times are simple duration times.

SUCCESSOR

PREDECESSOR		START 0	A 16	E 10	G 3	H 16	B 20	K 12	D 15	J 15	C 30	FINISH 0
START	0	0	0				0				0	
B	20	0		20					20			
C	30	0										30
H	16	35										16
A	16	0								16		
D	15	20			15	15						
J	15	16						15				
G	3	35						3				
E	10	20						10				
K	12	38										12
FINISH	0	51										

Fig. 9.8

Step 5

A forward pass is now carried out to find the *EST* of the various activities. This is done as follows:

(*a*) If the project can start NOW insert 0 in the START–START cell, otherwise insert the start date, calculated in elapsed time, from NOW as 0.

(*b*) Search the START row for any 'boxed' cells, and add the numbers within these cells to the START–START number. Enter the results in the START column beside the activity under which the cell had appeared (Fig. 9.9).

Step 6

Search across the rows that have just been labelled for boxed cells. For example in row *B*, which has just been labelled 0, there are two boxed cells, one in the *E* column, one in the *D* column. Since the rest of the columns under the headings *E* and *D* are vacant, the number in each of the boxes is added to the number in the START column, and the totals entered in the row titled with the same description as the column

PREDECESSOR \ SUCCESSOR		START 0	A 10	B 20	C 30
START	0	0	0	0	0
B	20	0			
C	30	0			
A	16	0			

Fig. 9.9

heading. Thus, the *B–E* cell contains a value 20 and as there are no other entries in the *E* column this is added to 0 (0 + 20 = 20) and this total placed in the START column against *E*. A similar calculation is carried out for the *B–D* cell and an entry made against *D* (Fig. 9.10).

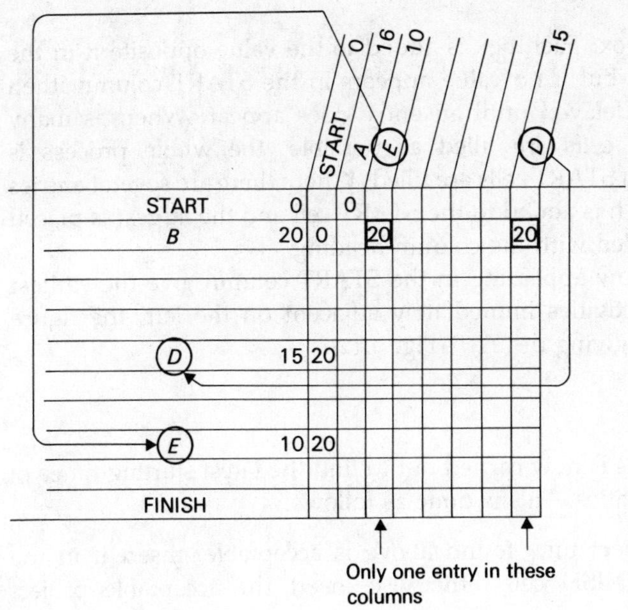

Fig. 9.10

Step 7

Once a boxed cell has been located it may be that the column containing that box holds other boxes, for example, the *C*–FINISH box (Fig. 9.11). Two other boxes are in that column, the *H*–FINISH box and

PREDECESSOR		SUCCESSOR: START 0	*A* 16	*E* 10	*G* 3	*H* 16	*B* 20	*K* 12	*D* 15	*J* 15	*C* 30	FINISH 0
START	0		0				0				0	
B	20			20					20			
C	30											30
H	16											16
A	16									16		
D	15				15	15						
J	15							15				
G	3							3				
E	10							10				
K	12											12
FINISH	0											

Fig. 9.11

the *K*–FINISH box. Each box is added to the value opposite it in the START column, but if no value appears in the START column, then this process is delayed until an entry does appear. When as many START column cells are filled as possible, the whole process is repeated until all START cells are filled. Where there are several entries in a column, each is added to the START cell and the largest is placed in the row labelled with the column heading.

The figures now appearing in the START column give the earliest times for the activities immediately adjacent on the left, the figure against FINISH giving the *TPT* (Fig. 9.12).

Step 8

A backward pass is now carried out to find the latest starting times of the various activities. This is done as follows:

(*a*) If the project time found above is acceptable, insert it in the FINISH–FINISH cell, otherwise insert the acceptable project time.

		SUCCESSOR										
		0	16	10	3	16	20	12	15	15	30	0
PREDECESSOR		START	A	E	G	H	B	K	D	J	C	FINISH
START	0	0	0				0				0	
B	20	0		20					20			
C	30	0										30
H	16											16
A	16	0								16		
D	15	20			15	15						
J	15	16						15				
G	3							3				
E	10	20						10				
K	12											12
FINISH	0											

Fig. 9.12

(*b*) Search the FINISH column for any 'boxed' cells and subtract their values from the FINISH-FINISH number. Enter these results in the FINISH row under the column heading that is the same as the 'boxed' row description. Thus, there is a boxed cell *K*–FINISH of value 12. Subtract this from the FINISH–FINISH value $(51 - 12 = 39)$. Enter this result under the *K* column. Similarly, enter $51 - 16 = 35$ under the *H* column and $51 - 30 = 21$ under the *C* column (Fig. 9.13).

(*c*) Using the new values in the FINISH row repeat the subtracting process. When a row contains more than one 'box' then all values must be subtracted from the FINISH row values, and the smallest entered under the appropriate column heading. Thus, referring to Fig. 9.13, the *K* column has three boxes (15, 3, 10) and none of these have boxes in the same rows, hence three entries (*a*) can be made:

a(i) From row *J* $39 - 15 = 24$ entered under *J*
a(ii) From row *G* $39 - 3 = 36$ entered under *G*
a(iii) From row *E* $39 - 10 = 29$ entered under *E*

There are now two boxes in row *D* each of which has a FINISH entry. Hence the entry to be placed under *D* is the smaller of:

$35 - 15 = 20$ or $36 - 15 = 21$

PREDECESSOR \ SUCCESSOR		START 0	A 16	E 10	G 3	H 16	B 20	K 12	D 15	J 15	C 30	FINISH 0
START	0	0	0				0				0	
B	20	0		20					20			
C	30	0										30
H	16	35										16
A	16	0								16		
D	15	20			15	15						
J	15	16						15				
G	3	35						3				
E	10	20						10				
K	12	38										12
FINISH	0	51				35		39			21	51

Fig. 9.13

and the value 20 is placed in the FINISH–*D* cell (entry *b*) (Fig. 9.14). Similarly, the *B* row has two boxes and the value entered under *B* is the smaller of:

$$29 - 20 = 9 \quad \text{or} \quad 20 - 20 = 0$$

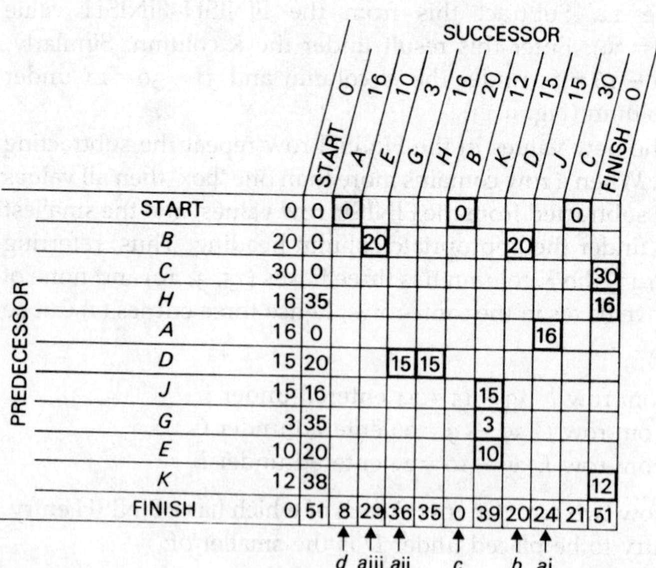

PREDECESSOR \ SUCCESSOR		START 0	A 16	E 10	G 3	H 16	B 20	K 12	D 15	J 15	C 30	FINISH 0
START	0	0	0				0				0	
B	20	0		20					20			
C	30	0										30
H	16	35										16
A	16	0								16		
D	15	20			15	15						
J	15	16						15				
G	3	35						3				
E	10	20						10				
K	12	38										12
FINISH	0	51	8	29	36	35	0	39	20	24	21	51
			↑ d	↑ aiii	↑ aii		↑ c		↑ b	↑ ai		

Fig. 9.14

that is, 0 is entered at FINISH–*B* (entry *c*). The final value (under *A*) is the difference between 24 and 16 = 8 and this is entered at FINISH–*A* (entry *d*).

The figures appearing in the FINISH row give the latest starting times for the activities at the column headings.

The matrix method, being effectively MoP, can treat with negative constraints. These are placed within the appropriate box, and ignored on the *forward* pass. On the backward pass they are subtracted from the START column and the difference compared with the difference between the positive value and the FINISH cell.

In practice it will be found that the matrix is easy to construct, and the subsequent analysis easier to perform than to describe. It is a configuration ideally suited for the construction of electrical analogues.

10 Analyzing the activity-on-node network: II Float

The forward and backward passes on activity-on-node (AoN) networking give the earliest start time (*EST*) and the latest start time (*LST*) for each activity. Thus, activity *J* (Fig. 10.1)

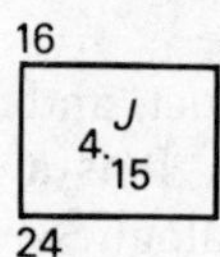

Fig. 10.1

may start at the beginning of week 16 or it may be delayed and start at the beginning of week 24 without increasing the total project time (*TPT*) of 51 weeks (Fig. 10.2).

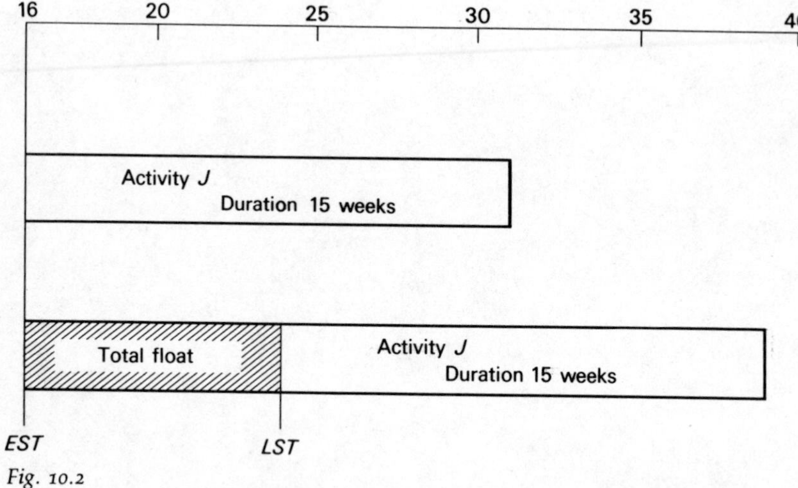

Fig. 10.2

Knowledge of this ability to 'move' or 'float' is valuable in many ways and the *total* amount an activity may move without affecting the *TPT* is called the *total float*.

> *Total float:* the total amount an activity may move without affecting the *TPT*.

Total float may be used to delay the start—and hence the finish—of an activity, or it may be used to increase the duration time of the activity, either or both of these being useful in deploying resources.

For activity *J* the total float is 8 weeks, given by:

Total float = latest starting time − earliest starting time

Since the two finishing times are given by adding a constant (the duration time) to the starting times:

Earliest finishing time = Earliest starting time + duration
Latest finishing time = Latest starting time + duration
EFT = *EST* + *d*
LFT = *LST* + *d*

then total float may be equally deduced from the finishing times:

Total float = latest finishing time − earliest finishing time

Total float, therefore, may be found readily by subtracting the upper figure on the box on either side from the lower figure on the lower box on the same side (Fig. 10.3).

6 10
Total float = 18 − 6 = 12 | *X* 4 | Total float = 22 − 10 = 12
18 22

Fig. 10.3

While total float does not affect the *TPT* it may delay the start of succeeding activities by 'shunting' them. One measure of the ability of an activity to float without affecting a subsequent activity is given by *free float*.

Free float: the float possessed by an activity which, if used, will not change the *TPT* nor change the float in any later activity.

Since free float involves consideration of later activities it cannot be calculated from the four start and finish times (*EST*, *LST*, *EFT* and *LFT*) of the activity. Reference must be made to all immediate successors, since it is the float in these that is to be unaffected.

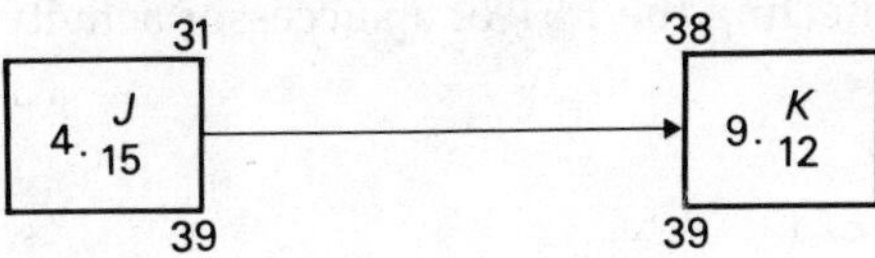

Fig. 10.4

Consider again activity *J*, this time with its immediate successor activity *K* (Fig. 10.4). If the total float in activity *J* of 8 weeks is used, then it will finish at week 39. This will then prevent activity *K* from starting at its *EST* of week 38—it will be 'pushed along' by 1 week, and its float will be diminished. Activity *J* may only float to week 38, the *EST* of its immediate successor, without affecting the float in that successor. Activity *J* may be said to have a *latest free finishing time.*

> *Latest free finishing time (LFFT):* the latest time by which an activity can finish without changing the *TPT* or changing the float in any later activity.

Free float is thus calculated from:

Free float = latest free finishing time − earliest finishing time

which, for activity *J* gives:

Free float *J* = 38 − 31 weeks = 7 weeks

When there are a number of succeeding activities it is necessary to scan them all to find the earliest of *all* the *EST*s, since this will give the *LFFT* of the activity being considered. In practice, this procedure is quite straightforward (Fig. 10.5).

Activity *L* has immediate successor activities *P*, *Q*, *R*, *S* and *T*, and all activities have other predecessors and successors not shown here. Forward and backward passes have been carried out giving:

EFT (L) = week 45
LFT (L) = week 64
EST (P) = week 73
EST (Q) = week 69
EST (R) = week 56
EST (S) = week 107
EST (T) = week 92

(*Note:* For the purpose of illustration all other *EST*s, *LST*s, *EFT*s and *LFT*s are suppressed as of no consequence in this discussion.)

The 'earliest' *EST* of all the succeeding activities is week 56. This is smaller than the *LFT* of the activity (*L*) being considered. Hence, *L* can only float to week 56 without affecting the *EST* of a successor activity, so that:

LFFT L = week 56
∴ Free float (*L*) = *LFFT* − *EFT*
= (56 − 45) weeks = 11 weeks

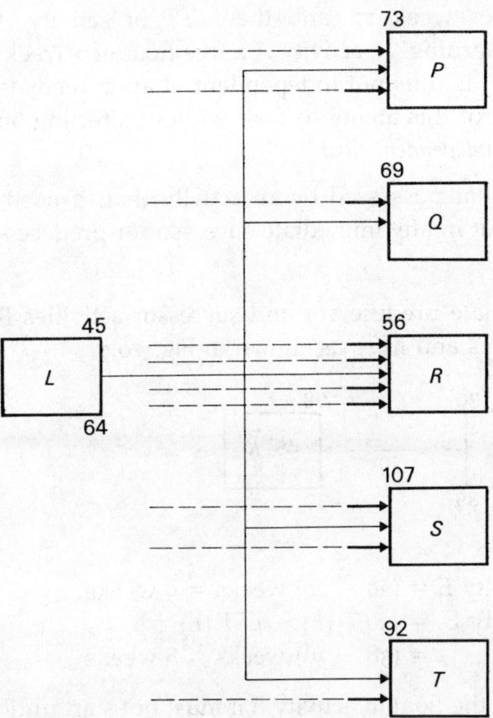

Fig. 10.5

With practice the calculation can be carried out even more simply without referring to the *LFFT*, although the author finds this a useful step when first considering free float.

Free float (X) = earliest *EST* of all immediate successors to X − *EST*(X)

Free float does not affect any immediate successor activities, but when an activity moves within its free float it may change the float condition in an immediate predecessor. Consider activity *J* with its immediate predecessor activity *A* (Fig. 10.6).

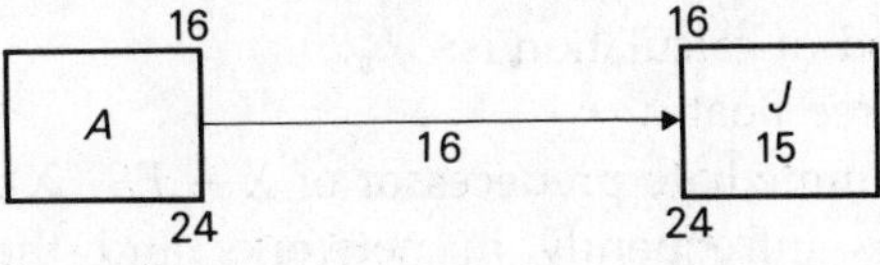

Fig. 10.6

It has already been shown that activity *J* has a free float of 7 weeks. Assume that its start is delayed by a small amount—say 1 week. Its starting time then becomes week 17, and the *LEFT* of activity *A* becomes week 17, thus 'awarding' to activity *A* a free float of 1 week. Any movement of activity *J* is thus not independent of an immediate predecessor. One measure of this ability to float without affecting an immediate predecessor is *independent float*.

> *Independent float:* the float possessed by an activity that, if used, will not change the float in any immediate successor or predecessor activity.

Activity *E*, with its immediate predecessor and successor activities *B* and *K*, has *EST*s, *LST*s, *EFT*s and *LFT*s as shown in Fig. 10.7.

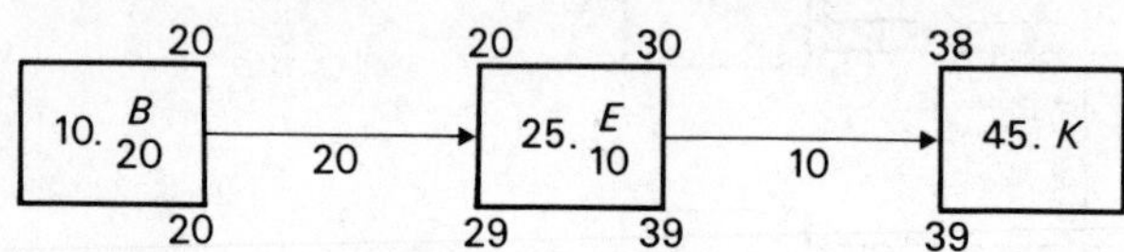

Fig. 10.7

The total float for activity E = (29 − 20) weeks = 9 weeks
The free float for activity E = $LFFT$ (E) − EFT (E)
= (38 − 30) weeks = 8 weeks

For activity *E* not to change the float in activity *B* it must not start until the latest finishing of activity *B* is reached. The *LFT* (*B*) is week 20. Hence, the earliest independent finishing time of activity *E* is (20 + 10) weeks = 30 weeks where:

> *Earliest independent finishing time (EIFT):* the earliest time an activity can finish without changing the *TPT* or changing the float in a previous activity.

Where there are a number of immediate predecessors it is necessary to scan them all to find the *latest* of all the *LFT*s. To this must then be added the duration time of the activity being considered to give the *EIFT*.

Notes:

(1) A possible simpler method of calculation is:
Independent float X = free float X −
(*LFT* of immediate predecessor of X − *EST* X).

(2) Independent float occurs infrequently in networks and the author has found very few fieldworkers who make use of it. Knowledge of total float *and ready reference to the network itself* is usually sufficient for project control.

These calculations, applied to the standard network

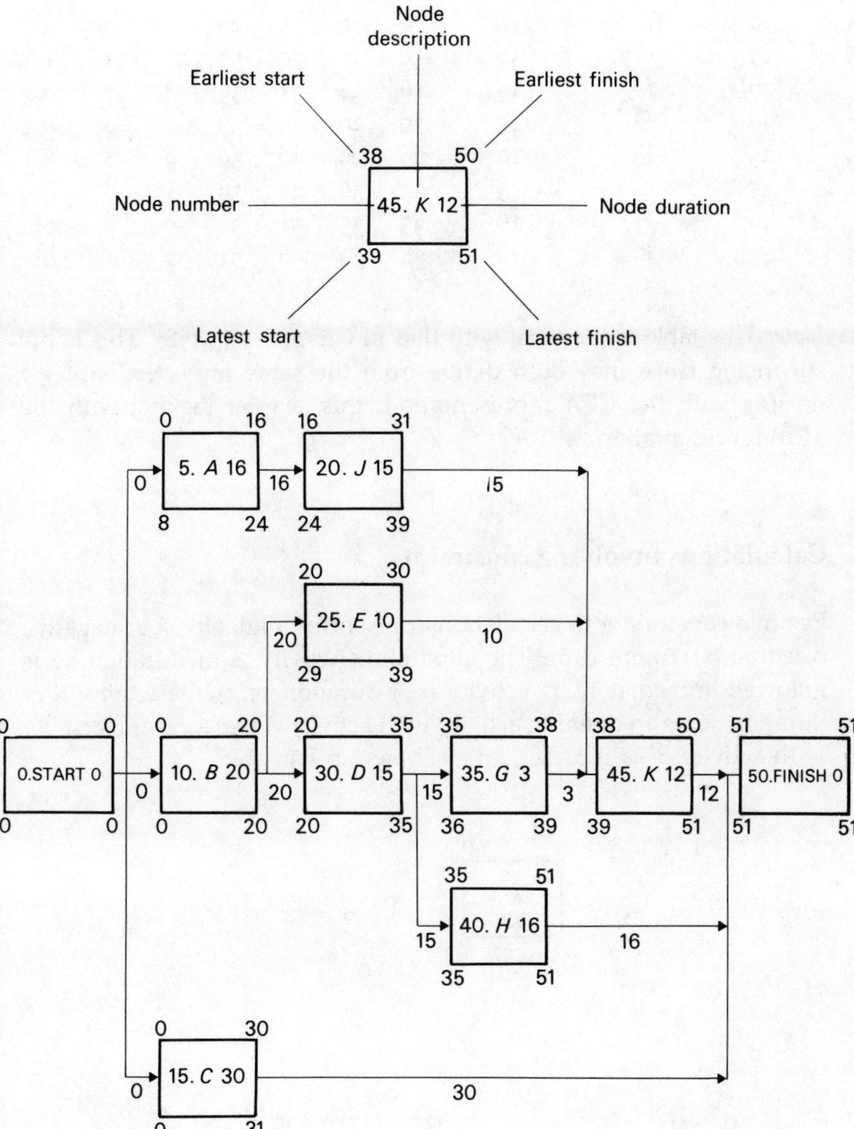

Fig. 10.8

will give the following:

	Activity		Start		Finish		Float		
No.	Description	Duration	Early	Late	Early	Late	Tot.	Free	Ind.
5	*A*	16	0	8	16	24	8	0	0
10	*B*	20	0	0	20	20	0	0	0
15	*C*	30	0	21	30	51	21	21	21
20	*J*	15	16	24	31	39	8	7	0
30	*D*	15	20	20	35	35	0	0	0
25	*E*	10	20	29	30	39	9	8	8
35	*G*	3	35	36	38	39	1	0	0
40	*H*	16	35	35	51	51	0	0	0
45	*K*	12	38	39	50	51	1	1	0

Note: This table is identical with that in Chapter 7 (p. 78). This is not surprising since they both derive from the same project, Chapter 7 dealing with the CPA representation, this chapter dealing with the MoP representation.

Calculations involving constraints

Positive constraints in calculations present no difficulty, but negative constraints require care. The situation: 'Activity *A* of duration 12 is followed immediately by activity *B* of duration 15, while activity *K* of duration 36, is an opening activity with activity *A* and a closing activity with activity *B'* is represented as shown in Fig. 10.9.

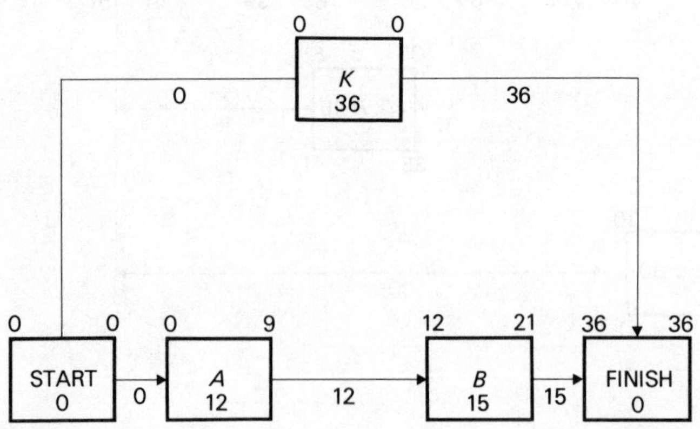

Fig. 10.9

The analyses for *A* and *B* are:

Activity		*Start*		*Finish*		
description	*Time*	*Early*	*Late*	*Early*	*Late*	*Float*
A	12	0	9	12	21	9
B	15	12	21	27	36	9

so that it would be possible to finish *A* at time 12 and start *B* at time 21, an interval of 9 time units between finishing *A* and starting *B*. This may not be acceptable: for some reason the interval between finishing *A* and starting *B* must be limited to, say, 4 units of time. A negative constraint of $-12 + (-4) = -16$ pointing backwards from *B* to *A* would result in Fig. 10.10.

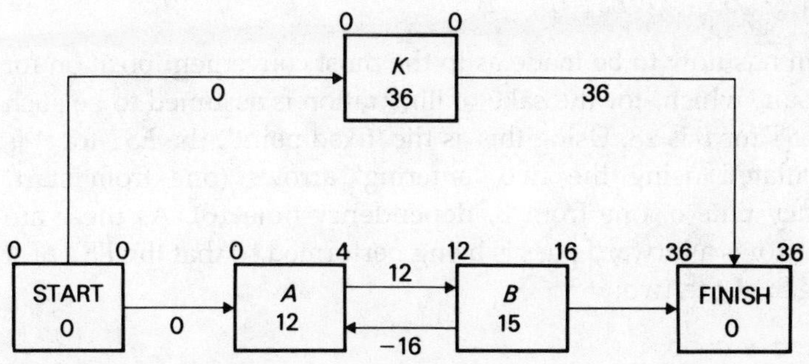

Fig. 10.10

The forward pass is carried out in the usual way by ignoring the negative constraint. On the backward pass, activity *B* has two emergent arrows (one, the negative constraint arrow). The figure for the latest start of *B* is determined by:

(1) Subtracting the normal (that is, non-negative) dependency time from the *LST* of its head activity:

$$36 - 15 = 21$$

and

(2) Subtracting the constraint dependency time from the *EST* of *its* head activity:

$$0 - (-16) = 16$$

and then choosing the smaller (16) from these two. The rest of the backward pass proceeds normally, and the resultant analysis for *A* and *B* is:

Activity description	*Duration*	*Start Early*	*Start Late*	*Finish Early*	*Finish Late*	*Float*
A	12	0	4	12	16	4
B	15	12	16	27	31	4

Thus, even with *A* finishing as early as possible (12) and *B* starting as late as possible (16), the 'gap' between *A* and *B* cannot exceed 4.

In effect this 'ties' *A* and *B* to their earliest possible positions. It may be that a later position is more appropriate. In this case the procedure is to determine the earliest start of *A* and the latest start of *B*, *ignoring the negative constraint* which in this example would give:

EST (*A*) 0: *LST* (*B*) 21

A decision has now to be made as to the most convenient position for the *A–B* pair, which, for the sake of illustration is assumed to be such that the *LST* for *B* is 18. Using this as the 'fixed point', the *EST* for *A* is now calculated using the two 'entering' arrows (one from start, dependency time 0, one from *B*, dependency time 16). As these are entering arrows a forward pass is being performed so that the *EST* of *A* is the larger of the two:

$$0 + 0 = 0$$

or

$$18 + (-16) = 2$$

that is the *EST* of *A* is 2. This is used in place of the previously obtained figure of 0, which is struck out and the *EST* of *B* (2 + 12 = 14) and the *LST* of *A* (18 − 12 = 6) are redetermined (Fig. 10.11).

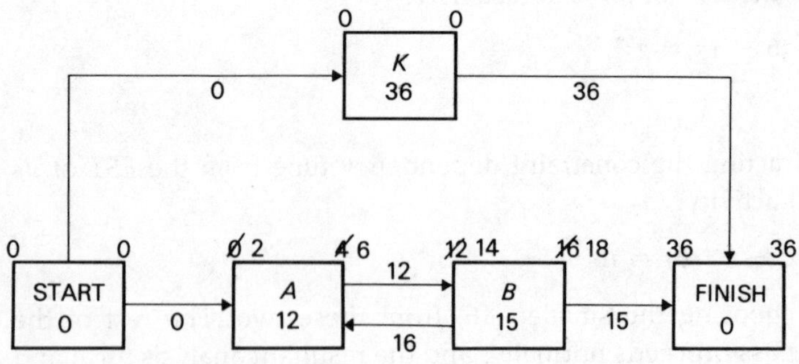

Fig. 10.11

Paired jobs

There are circumstances when one job must start *immediately* after its predecessor has been completed. In MoP this is represented by the diagram in Fig. 10.12.

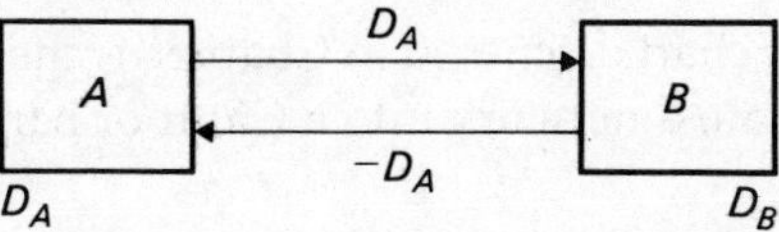

Fig. 10.12

A possibly simpler procedure is to combine the two activities into one activity ($A + B$) with a duration time ($D_A + D_B$).

Note: All the above calculations apply to simple, finish-to-start dependencies. Multiple dependencies follow a very similar pattern; float calculations may be made by reference back to the fundamental definitions of float.

11 The network and the Gantt chart

Despite the difficulties with the Gantt chart discussed in Chapter 1, the author has found the ability to translate a network into a Gantt or bar chart valuable for four reasons:

(1) It is more readily understood by 'unskilled' persons at all levels than an arrow diagram. The author knows one company where all plans are made using the project network techniques (PNT) but they are presented to top management as bar charts.
(2) Progress can easily be displayed on it.
(3) The meaning of float and activity times of all kinds can be understood by examining a bar chart. The author has found the drawing of a bar chart an invaluable aid when teaching.
(4) Simple resource allocation can be performed on, and the mechanics of complex resource allocation illuminated by consideration of a bar chart.

There are, broadly, two ways whereby a bar chart may be derived from a network.

(1) By carrying out a forward and backward pass, calculating the various activity times (either manually or by computer) and using these to give the locations of the time-scaled activity bars. This is so simple that it requires no explanation. It is the only way in which part of a network can be drawn as a bar chart.
(2) By drawing the activities to a time-scale and by using the node number to ensure correct inter-relationships. There are a variety of methods of carrying out this transformation. Over the years the author has come to the conclusion that the *time-scaled network* is the most useful form.

The time-scaled network

The network that has been used extensively earlier will be employed again here as an illustration (Fig. 11.1). Any network, particularly if it has been drawn according to the 'activities parallel to the edges of the paper' recommendation can be readily time-scaled.

1. Activity-on-arrow (AoA)

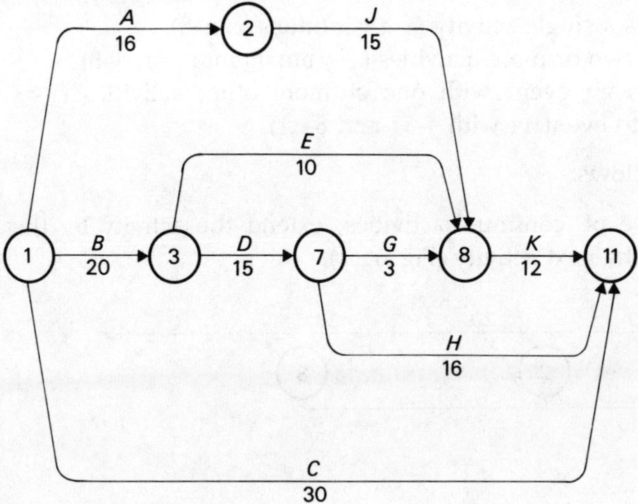

Fig. 11.1

Start at initial event and draw to scale all opening activities, identifying them by their event numbers. It is desirable at this stage to space these well out on the page (Fig. 11.2).

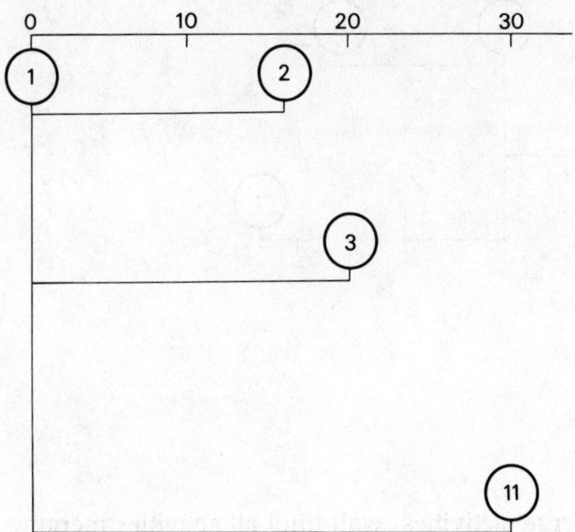

Fig. 11.2

Each activity will then either:

(*a*) *Continue* as a single activity (1–2 continues as 2–8).
(*b*) *Burst* into two or more activities (1–3 bursts into 3–7, 3–8).
(*c*) *Merge* into an event with one or more other activities (1–11 merges into event 11 with 7–11 and 8–11).

Proceed as follows:

(*a*) In the case of 'continue' activities, extend the activity by the length of the next activity (Fig. 11.3).

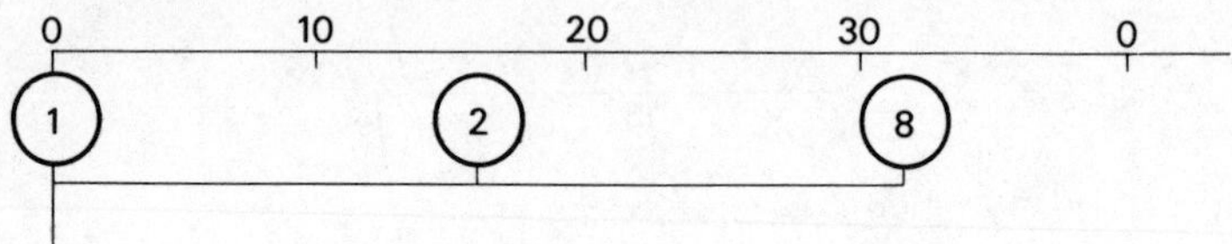

Fig. 11.3

(*b*) In the case of 'burst' activities, draw a single vertical line and from this draw to scale the 'bursting' activities (Fig. 11.4).

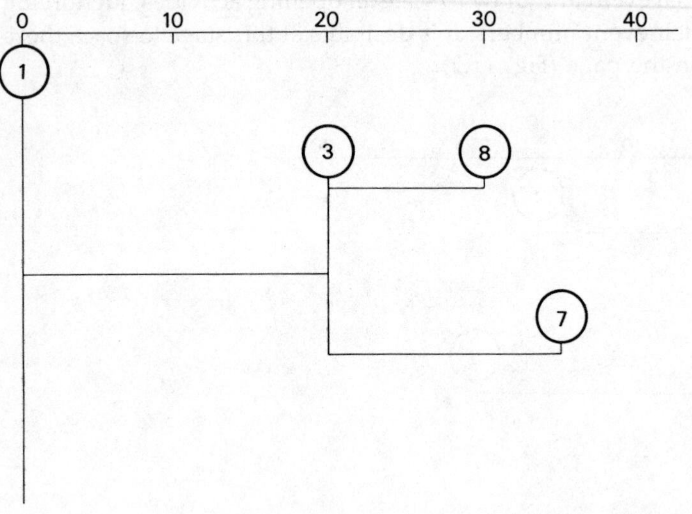

Fig. 11.4

(*c*) In the case of 'merge' activities, wait until all activities merging into a common event have been drawn—for example, wait until activities 2–8, 3–8 and 7–8 have been drawn, and then draw a vertical line to form a 'barrier' across the end of the activity that extends furthest to the right. Join all the merge activities to that fence by means of dotted lines. These lines represent free float (Fig. 11.5).

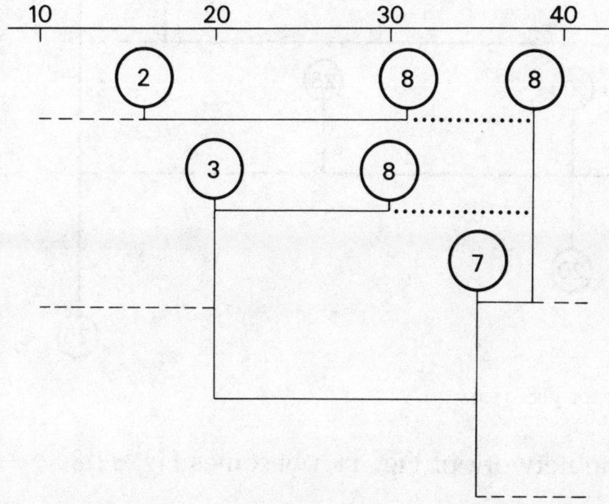

Fig. 11.5

Repeat the above until the last event is reached. Re-draw if desired to emphasize special organizational or resource features.

Note: Any dummy must be drawn in as a vertical line, its head and tail numbers being shown. For example, Fig. 11.6

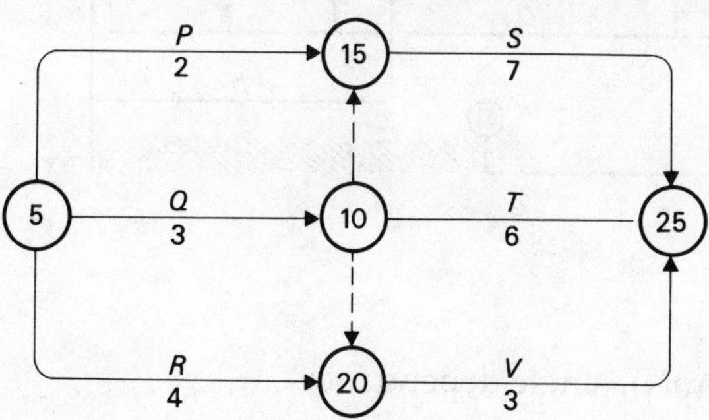

Fig. 11.6

would become Fig. 11.7.

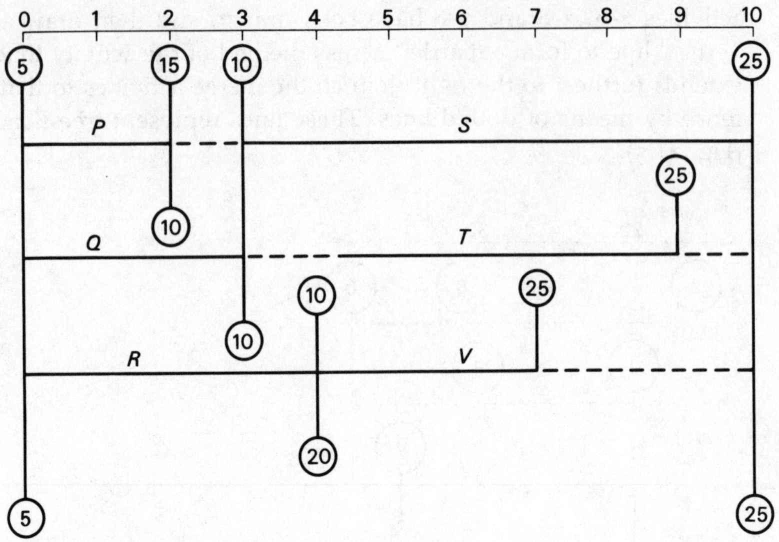

Fig. 11.7 Time-scaled network for Fig. 11.6

Using this procedure, the network of Fig. 11.1 becomes Fig. 11.8.

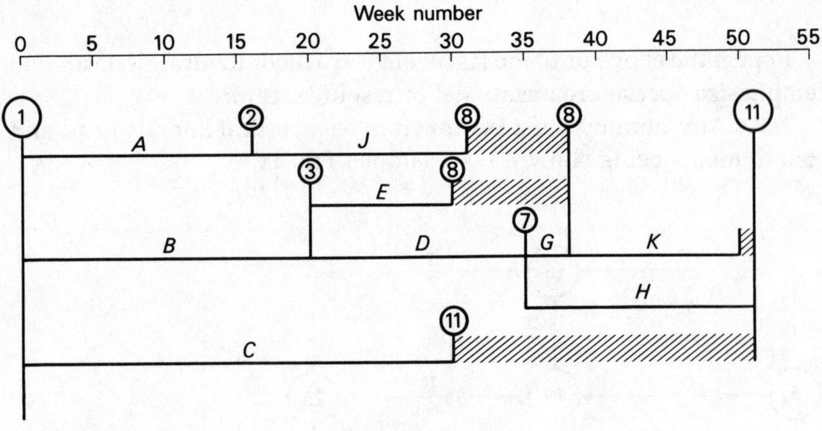

Fig. 11.8

2. Activity-on-node (AoN), single dependency

Start at the initial activity START and draw this as a vertical line (Fig. 11.9). Draw to scale all activities *A*, *B* and *C*. Draw a short vertical line at the end of each of these activities. Place the activity numbers at the

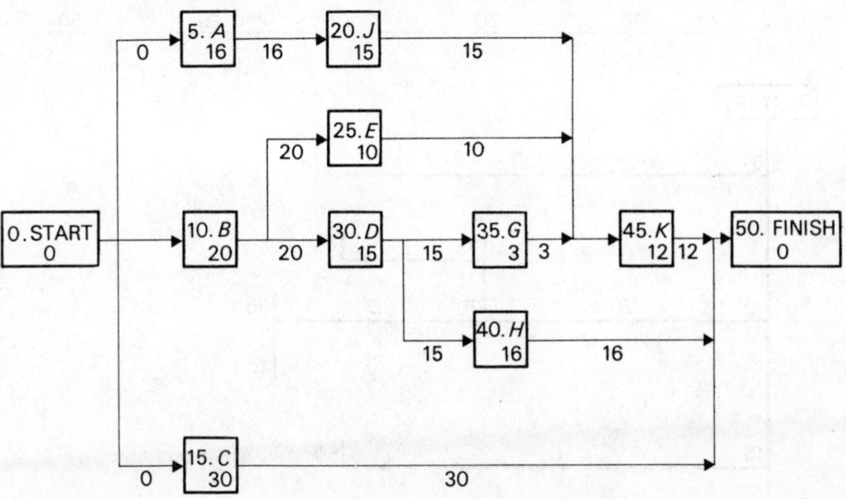

Fig. 11.9

beginning of each time-scaled activity. To the *right* of the vertical end-line inscribe the number(s) of the appropriate succeeding activities (Fig. 11.10).

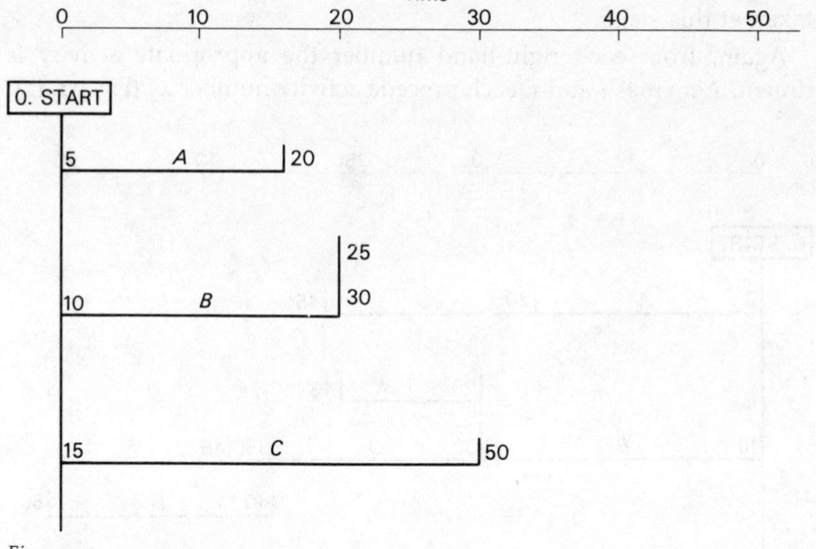

Fig. 11.10

From each right-hand number draw the appropriate activity forward to scale (Fig. 11.11). Thus, from '20' draw activity *J*, duration 15. After

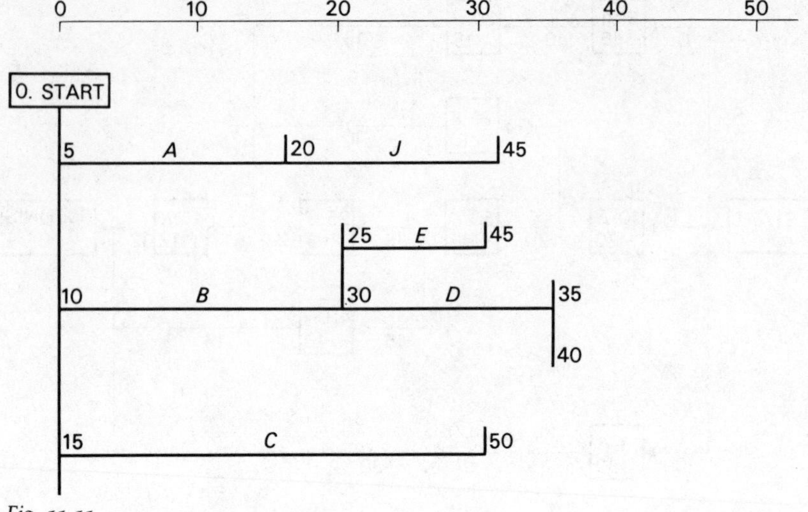

Fig. 11.11

activity *B* there are two numbers 25 and 30. Accordingly, two activities need to be drawn (*E* and *D*). Activity *C* precedes activity FINISH, but since other activities not yet drawn also precede FINISH, no action is taken at this stage.

Again, from each right-hand number the appropriate activity is drawn. Activities *J* and *E* each precede activity number 45 (Fig. 11.12).

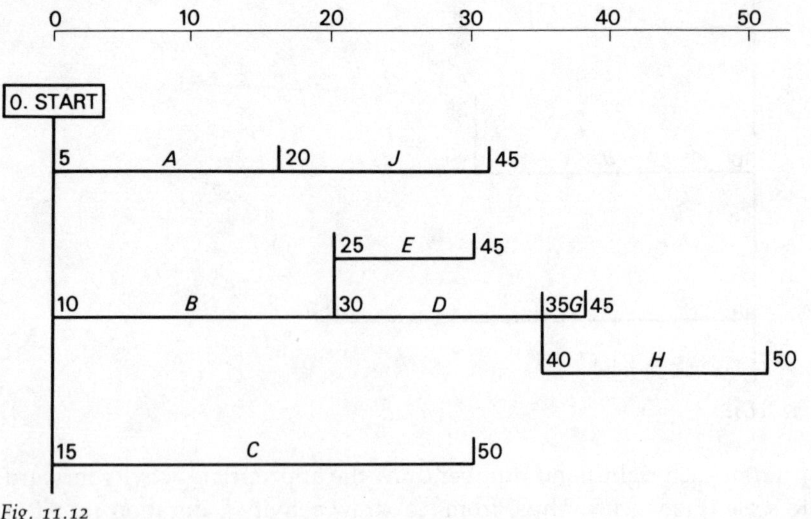

Fig. 11.12

Examination of the network shows that three activities *J*, *E* and *G* precede this, and as only two have been drawn, no drawing takes place until this third predecessor is in place. However, activities *G* and *H* can be drawn; the three activities preceding activity number 45 (*K*) having now been drawn, it can be located. Clearly it cannot start until the latest completion date of all predecessors, that is, after *G*. A 'fence' drawn up from the start of activity *K* enables the floats of *E* and *J* to be drawn. Activity *K* also precedes activity number 50 ('FINISH') as does activity *H*. Accordingly, this activity too can be drawn in, completing the diagram (Fig. 11.13).

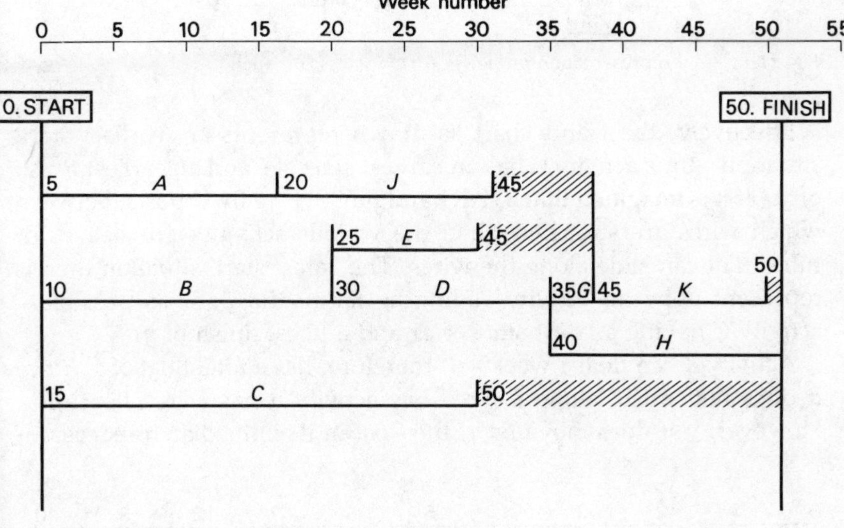

Fig. 11.13

The network of Fig. 11.14—the AoN representation of the AoA network of Fig. 11.6—translates very simply into a time-scaled network using the above procedure (Fig. 11.15).

Analysis by Gantt chart

The diagrams of Figs 11.8 and 11.13, one derived from the AoA network, the other from the AoN representation, give, not surprisingly, the same diagram. The analyses of Chapters 6, 7, 9 and 10 can be carried out from this diagram.

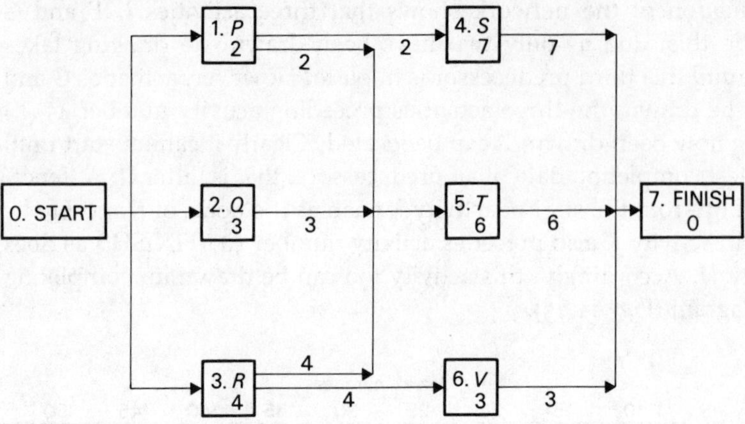

Fig. 11.14 AoN representation of Fig. 11.6

Effectively, the Gantt chart as drawn represents an 'earliest start' situation—thus activity *C* has an earliest start of 0 and an earliest finish of 30. If it is imagined that START and FINISH are fixed posts, between which wires are stretched, then the various activities are sleeves or tubes that can slide along the wires. The 'latest start' situation then is represented by an activity sliding as far to the right as possible—activity *C* having a latest start of 21 and a latest finish of 51.

Activity *K* can float 1 week—it, therefore, has a total float of 1. Since it does not 'push' along a *succeeding* activity it has a free float of 1. However, if it does move by 1, the float in its immediate predecessor

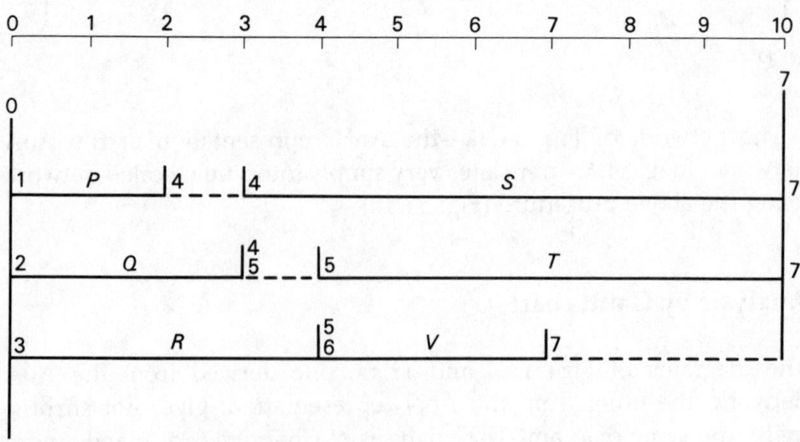

Fig. 11.15 Time-scaled network for Fig. 11.14. Compare with Fig. 11.7

changes—*it* can float freely. Hence, activity *K* has no independent float since any movement affects a predecessor.

Activity *G* can 'move' by 1, but only by 'pushing' activity *K* forward—hence it has a total float of 1 and a free float of 0.

Activity *J* has a total float of 8—a movement of 7 before it 'pushes' activity *K* and a further 1 as it shunts activity *K* along—hence, it has a total float of 8 and a free float of 7. It has no independent float since it 'opens up' the float situation for activity *A*.

All the results of the table on pp. 78 and 110 are deducible in the above fashion, and students are strongly advised to do so. In the early days of studying PNT it will often be found that calculation difficulties can be resolved by cross-checking with a Gantt chart.

12 Precedence networks—multiple dependency AoN

The activity-on-node (AoN) system so far described sets out essentially only one relationship between activities: the start of an activity depends upon the starts of its predecessors. In the 1960s, IBM developed the 'System 360 Project Control System', which used multiple dependencies and for this the name 'precedence network' or 'precedence diagram' was used. This was an unhappy terminology in the sense that several years earlier the name 'precedence diagram' was given to the diagram setting out the necessary precedence relationships in a production flow line. However, the name is now blessed by the BSI and is widely used in AoN literature, and accordingly it is used here.

There are four dependencies that can be described:

(1) *Finish-to-start* (or *normal*) (Fig. 12.1)

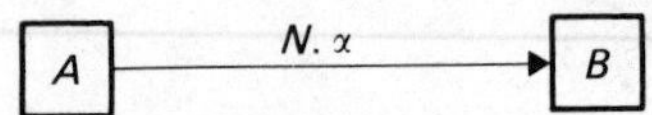

Fig. 12.1 Finish-to-start dependency

Activity *B* may not start until at least α days after the finish of activity *A*. The *N* is the most common coding applied to this arrow and indicates the relationship. It is entered into the input data in a computer system. If activity *B* may follow immediately upon activity *A*, then the α becomes 0 and is either so written or it is ignored.

(2) *Start-to-start* (Fig. 12.2)

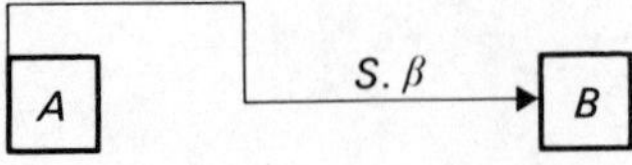

Fig. 12.2 Start-to-start dependency

Here, at least β days must elapse between the start of activity *B* and the start of activity *A*. The situation where $\beta > 0$—when the start of activity *B* lags behind the start of activity *A*—is sometimes called a 'lag-start' relationship. Again the S represents a common coding, and again if $\beta = 0$ it is either so written or ignored.

(3) *Finish-to-finish* (Fig. 12.3)

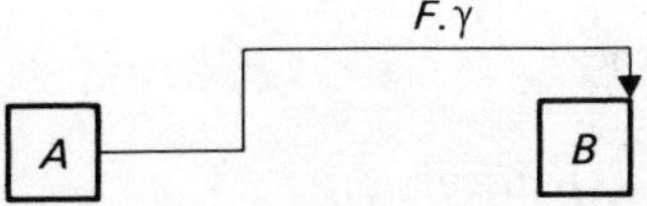

Fig. 12.3 Finish-to-finish dependency

Here, at least γ days must elapse between the completion of activity A and the completion of activity B. When $\gamma > 0$ the relationship is said to be a 'lag-finish' relationship. The usual coding is F, and a zero lag is either written 0 or ignored.

(4) *Parallel* (Fig. 12.4)

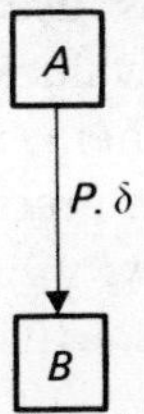

Fig. 12.4 Parallel dependency

Activity B may start δ days after the start of activity A but it may not finish after the finish of activity A.

The lag-start and lag-finish dependencies may be combined (Fig. 12.5).

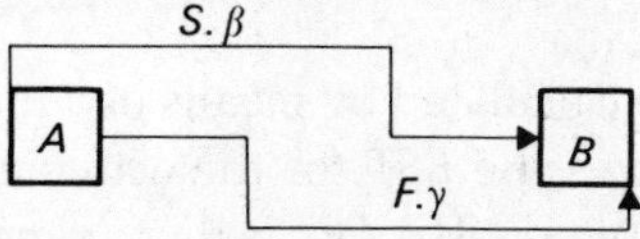

Fig. 12.5 Lag-start, lag-finish dependencies

Here, activity B may not start until β days after the start of activity A, and cannot finish until γ days after the finish of activity A. This is the 'ladder' situation of activity-on-arrow (AoA) and Chapter 8 should be read to understand the resource implications of this situation. This lag-start, lag-finish form may be used to represent the parallel dependency (4) above by setting $\gamma = 0$ (Fig. 12.6).

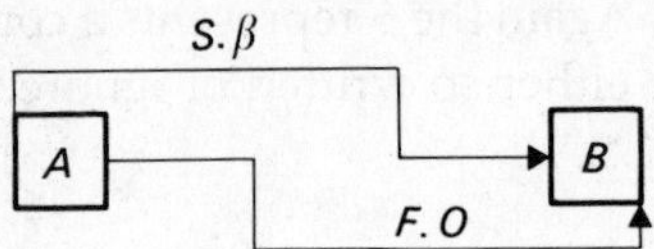

Fig. 12.6 Alternative parallel dependency

This form is to be preferred to (4) since:

(*a*) It minimizes the number of symbols.
(*b*) It ensures, when carrying out manual calculations, that constraints apply at *both* starts and finishes. This may be overlooked with the single arrow of (4).

Notes:

(*a*) α, β, γ and δ may be positive (indicating a delay), zero or negative (indicating an overlap). If a computer is used for calculations it is important to discover if the negative values may be used as not all computer programs can handle them. Some workers write α, β, γ and δ as percentages of the preceding activity's duration—thus, if the duration time of activity *A* is 10 days and activity *B* may start 2 days after the start of activity *A*, then α would be written: 20 per cent.
(*b*) Some workers define other relationships, for example a 'start-to-finish' dependency. The author has yet to see a 'middle-to-middle' relationship, but doubtless it is being used somewhere!

Activity times and precedence networks

Earliest start activity times (*EST*s) are calculated by means of a forward pass, where progressive addition gives the *EST* for the activities and the total project time (*TPT*) for the network. The *TPT* is then (generally) used to set the latest finishing time (*LFT*) for the network as a whole, and a backward pass, with successive subtraction, gives the latest finish activity times.

A tabular system is often recommended for manual calculation; the author finds that a manual calculation *on the network itself* is simpler and gives a more useful insight into the physical meaning of the network and its constituent parts. The method is illustrated by a series of examples in which the time units are consistent and therefore not specified.

Finish-to-start (*N*)

Three activities *A*, *B* and *C* with durations 5, 10 and 15 precede an activity *X* (duration 20). *X* may not start until at least 3 after the completion of *A*, at least 4 after the completion of *C*, but may start immediately after the completion of *B*. The *EST*s of *A*, *B* and *C* are 10, 12 and 14 respectively—derived from some other portion of the network not represented here (Fig. 12.7).

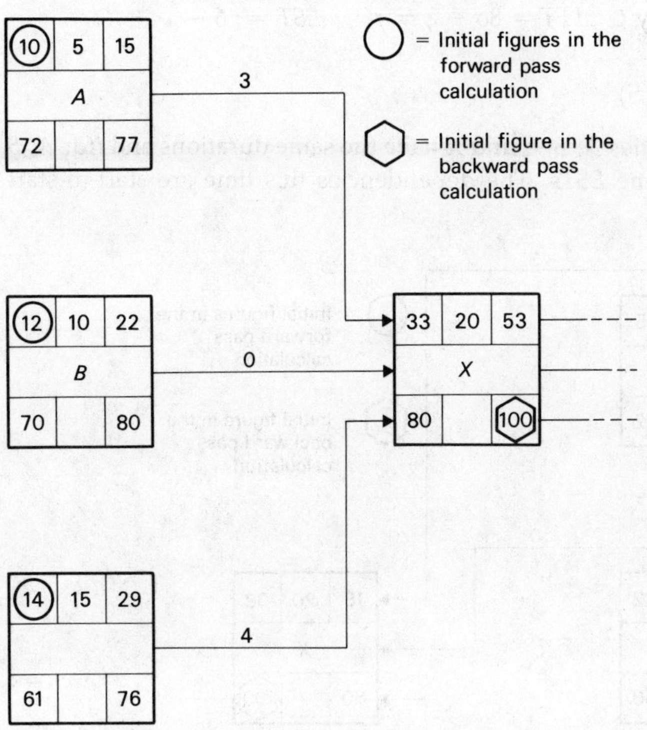

Fig. 12.7 Finish-to-start calculations

The earliest finishing times (*EFT*s) of the three activities are:

$EFT\ A = 10 + 5 = 15$
$EFT\ B = 12 + 10 = 22$
$EFT\ C = 14 + 15 = 29$

The *EST* for activity *X* is the time when the preceding *EST*s and their dependency times are all complete, that is, it is the largest of:

$15 + 3 = 18; \quad 22 + 0 = 22; \quad 29 + 4 = 33$

that is, 33. Since the duration of activity *X* is 20, its *EFT* is 53, and this is filled in on the forward pass for subsequent calculations.

The *LFT* for activity *X*, derived from some other part of the network is 100, consequently its *LST* is 80. Since 3 must elapse between the finish of activity *X* and the start of activity *A* then its *LFT* must be $80 - 3 = 77$ and its *LST* is $77 - 5 = 72$.

Similar calculations give:

For activity *B* $LFT = 80 - 0 = 80$: $LST = 80 - 10 = 70$
For activity *C* $LFT = 80 - 4 = 76$: $LST = 76 - 15 = 61$

Start-to-start (*S*)

The four activities *A*, *B*, *C* and *X* have the same durations and (for *A*, *B* and *C*) the same *EST*s. The dependencies this time are start-to-start dependencies.

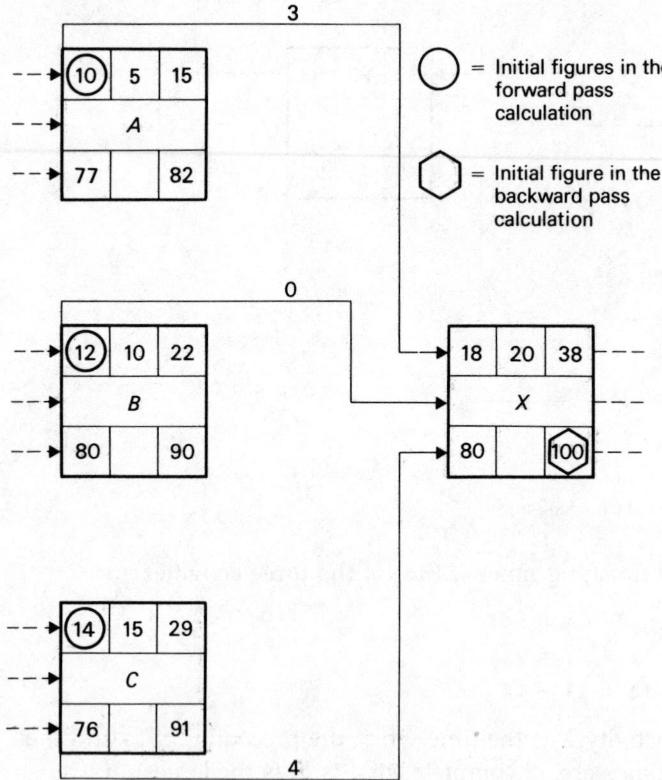

Fig. 12.8 Start-to-start calculations

The *EST* of *X* is the latest of:

$$10 + 3 = 13; \quad 12 + 0 = 12; \quad 14 + 4 = 18;$$

that is, 18. The *EFT*s are then:

$$\begin{aligned} EFT\ (A) &= 10 + \ \ 5 = 15 \\ EFT\ (B) &= 12 + 10 = 22 \\ EFT\ (C) &= 14 + 15 = 29 \\ EFT\ (X) &= 18 + 20 = 38 \end{aligned}$$

It is given, from some other part of the network, that the *LFT* for activity *X* is 100, hence:

$$LST\ (X) = 100 - 20 = 80$$

Since activity *X* may not start until 3 units of time have elapsed after the start of activity *A*, then:

$$\begin{aligned} & LST\ (A) = 80 - 3 = 77 \\ \text{Similarly} \quad & LST\ (B) = 80 - 0 = 80 \\ \text{and} \quad & LST\ (C) = 80 - 4 = 76 \end{aligned}$$

and hence (Fig. 12.8):

$$\begin{aligned} LFT\ (A) &= 77 + \ \ 5 = 82 \\ LFT\ (B) &= 80 + 10 = 90 \\ LST\ (C) &= 76 + 15 = 91 \end{aligned}$$

Finish-to-finish (*F*)

The same four activities are used in illustration as before with an additional activity *Y*, duration 25, which has a finish-to-finish relationship of 4 with activity *C* and an *LFT* of 90.

The *EFT*s of activities *A*, *B* and *C* are:

$$\begin{aligned} EFT\ (A) &= 10 + \ \ 5 = 15 \\ EFT\ (B) &= 12 + 10 = 22 \\ EFT\ (C) &= 14 + 15 = 29 \end{aligned}$$

Activity *X* cannot finish until 3, 0 and 4 time units have elapsed after the finish of its predecessors, hence the *EFT* of *X* is the latest of:

$$15 + 3 = 18; \quad 22 + 0 = 22; \quad 29 + 4 = 33$$

that is:

$$EFT\ (X) = 33$$

hence: $EST\ (X) = 33 - 20 = 13$

(*Note:* If another dependency between *X* and a predecessor—say a start-to-start dependency—had 'required' *X* to have a *later EST*, then this would have been adopted, and the resulting *later EFT* also used.)

Since activity *X* may not finish until 3 time units have elapsed after the completion of activity *A*, and the *LFT* of *X* (determined elsewhere) is 100, then the *LFT* of *A* is 100 − 3 = 97. Equally:

$$LFT\ (B) = 100 - 0 = 100$$

Activity *C* has two finish-to-finish relationships (Fig. 12.9) and

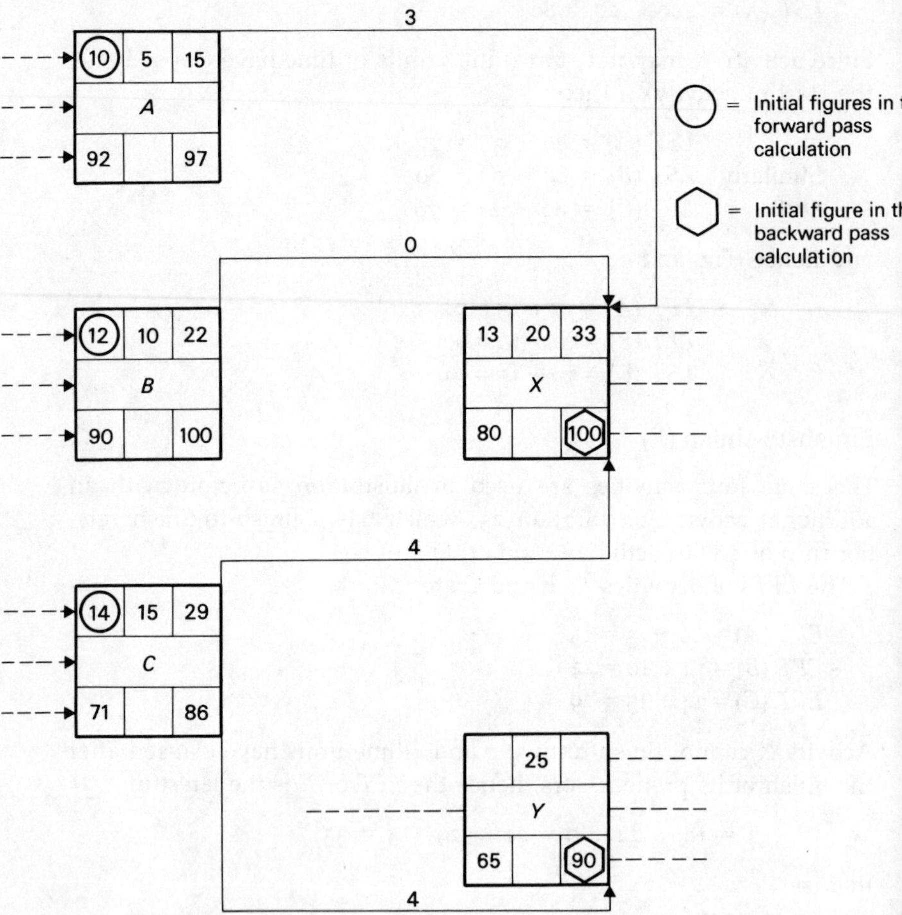

Fig. 12.9 Finish-to-finish calculations

consequently its *LFT* is the *earlier* of:

100 − 4 = 96 (from activity *X*)
and 96 − 4 = 86 (from activity *Y*)
that is: *LFT* (*C*) = 86

*EFT*s are derived by subtracting duration times from *LFT*s.

Several dependencies at a node

The fact that two 'entering' dependencies may have arrow-heads at different ends of a node may cause one or other to be overlooked in manual calculation. For example, activity *L* (duration 10) has a finish-to-start dependency of 1 with activity *J* (duration 2, *EST* 7) and a finish-to-finish relationship of 6 with activity *K* (duration 12, *EST* 13). The activity starting times for *L* are calculated:

EST—the earlier of 9 + 1 = 10 and 25 + 6 − 10 = 21
that is, 10

LST—the later of 9 + 1 + 10 = 20 and 25 + 6 = 31
that is, 31

The *LFT* for *L*, determined elsewhere, is 45, hence (Fig. 12.10):

EFT (*L*) = 45 − 10 = 35
LFT (*K*) = 45 − 6 = 39: *EFT* (*K*) = 39 − 13 = 26
LFT (*J*) = 45 − 1 = 44: *EFT* (*J*) = 44 − 2 = 42

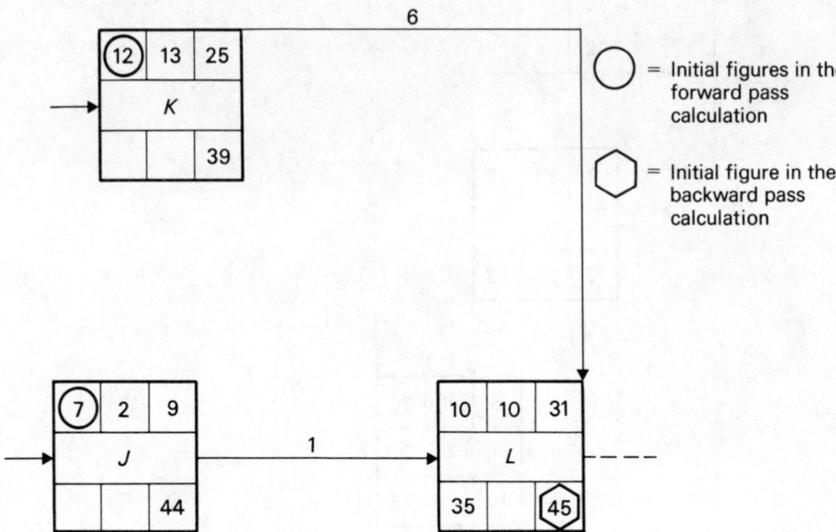

Fig. 12.10

Lag-start, lag-finish

Four activities *A*, *B*, *C* and *D* (durations 6, 18, 10 and 8 respectively) are in parallel, with the following constraints:

Activity *B* may not start until at least 1 time unit after the start of activity *A*. At least 2 time units are required for the completion of activity *B* after activity *A* is finished.

Activity C may not start until at least 2 units of time after the start of activity *B*. At least 3 time units are required for the completion of activity C after activity *B* is finished.

Activity *D* may not start until at least 2 units of time after the start of activity C. At least 4 units of time are required for the completion of activity *D* after activity C is finished.

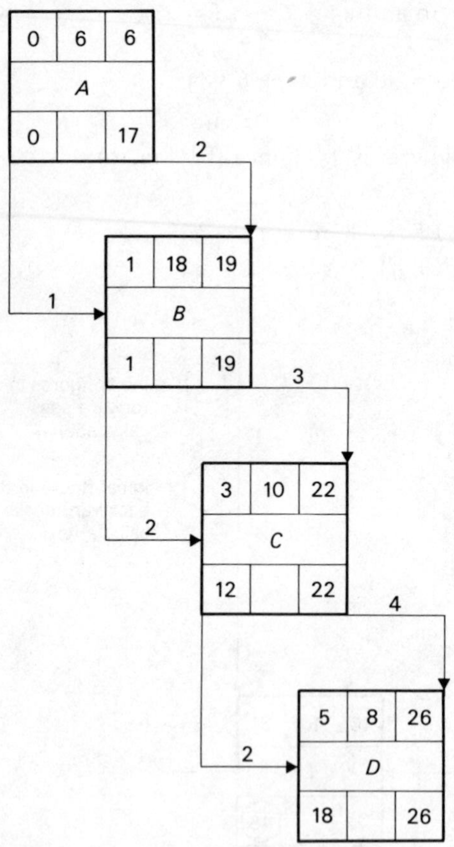

Fig. 12.11 Lag-start, lag-finish activities

This situation is represented by Fig. 12.11. Assume that the *EST* for activity *A* is 0, then the *EFT* for activity *A* is 0 + 6 = 6.

Assume that the *EST* for activity *B* is 0 + 1 = 1, then the *EFT* for activity *B* is the later of:

6 + 2 = 8 and 1 + 18 = 19, that is, 19

the *EST* for activity *C* is 1 + 2 = 3, then the *EFT* for activity C is the later of:

3 + 10 = 13 or 19 + 3 = 22, that is, 22

the *EST* for activity *D* is 3 + 2 = 5, then the *EFT* for activity *D* is the later of:

5 + 8 = 13 or 22 + 4 = 26, that is, 26

Assume that activity *D* is a closing activity, then the *LFT* for activity *D* is set at 26.

(*Note:* If other activities are present they may set a larger *LFT* to be adopted.)

The *LST* for activity *D* is 26 − 8 = 18.

The *LFT* for activity C is 26 − 4 = 22, then the *LST* for activity C is the earlier of:

22 − 10 = 12 or 18 − 2 = 16, that is, 12

The *LFT* for activity *B* is 22 − 3 = 19, and the *LST* for activity *B* is the earlier of:

19 − 18 = 1 or 12 − 2 = 10, that is, 1

The *LFT* for activity *A* is 19 − = 17, and the *LST* for activity *A* is the earlier of:

17 − 6 = 11 or 1 − 1 = 0, that is, 0

Note: the above situation is that discussed in Chapter 8 where the resource implications—which are substantial—are discussed. Readers are advised to study Chapter 8 if they have not already done so.

Float

Float is defined as the time available for an activity in addition to its duration. For an activity *N*, duration *d*:

Time available = *LFT* − *EST*
∴ Float = *LFT* − *EST* − *d*

Thus, for activity *L* of Fig. 12.10:

$$\text{Float} = 45 - 10 - 10 = 25$$

This calculation assumes that all previous activities start as early as possible, and all succeeding activities finish as late as possible (Fig. 12.12), hence this is the *total* amount of float – the *total float*:

Total float = latest finishing time – earliest starting time – duration

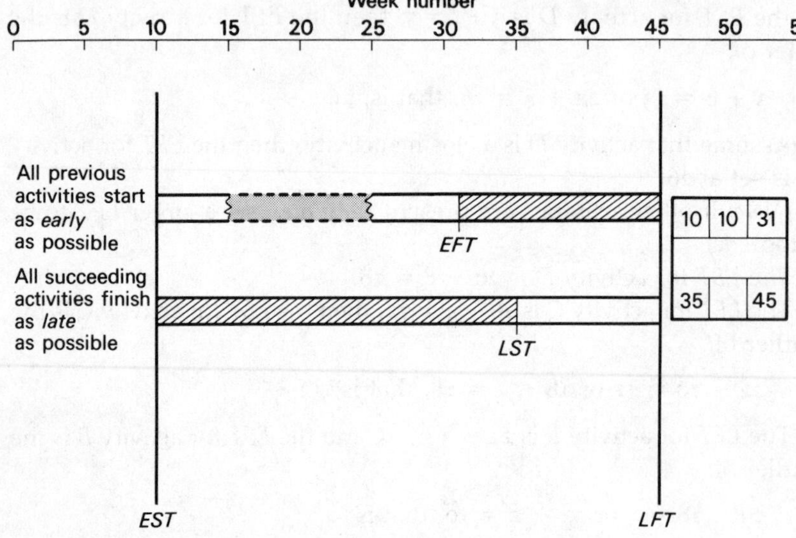

Fig. 12.12 Total float in precedence diagrams

Note: The presence of multiple dependencies at the node may make the calculation used in single dependency AoN:

$$\begin{aligned}\text{Total float} &= LST - EST \\ &= LFT - EFT\end{aligned}$$

invalid since there may be a starting or finishing constraint on an early or late start or finish. Thus, for example, the finish-to-finish relationship between activities *K* and *L* means that even though activity *L* has an *EST* of 10, it has an *EFT* of 31. Some workers define:

$$\text{Early total float} = LFT - EFT$$
$$\text{Late total float} = LST - EST$$

but these terms are not in general use.

In AoA and single dependency AoN two other floats—free float and independent float—are identified. The presence of multiple dependencies of different kinds at a node means these types of float cannot, generally, be calculated, and in precedence networking only total float is calculated and used.

Note: In practice, most fieldworkers, whatever type of networking is being used, find that total float provides all the information necessary, hence the loss of the ability to calculate anything but total float is of little practical use.

Node symbols in Activity on Node Networking

The latest version of the British Standard BS4335 makes the following recommendations concerning the symbols to be used for a Node in Activity-on-Node Networking.

Earliest start time	Duration	Earliest finish time
Activity number Activity description Resources required		
Latest start time	Total float	Latest finish time

Fig. 12.13 Single Dependency Networking

Earliest start time	Latest start time
Activity number Actiyity description Resources required	
Activity duration	Total float

Fig. 12.14 Method-of-potentials Networking

Activity description Resources required	
Activity number	Activity duration

Fig. 12.15 Multiple Dependency Networking

It will be noted that the first two symbols permit manual calculations to be written in the node box on the Network. The likelihood of such manual calculations being carried out in multiple dependency networking is so small that no provision is made in the Node box for writing in such calculations.

13 Reducing the project time

As already stressed, a network is a statement of policy, that is, a statement of the means whereby an objective is to be obtained. It is extremely rare for only one acceptable policy to be formulated; further, it is equally true that almost any policy can be improved. This is very clearly recognized by the work study engineer, whose basic tenet of faith is 'There is always a better way'.

This being so, it is desirable that when a network has been drawn it should be very carefully and critically examined. In complex projects, the difficulty in examining all activities is so great that frequently no examination takes place at all, or alternatively only those activities in which the examiner has a particular interest are inspected. Project network techniques (PNT) have the tremendous advantage that, by isolating the *critical* activities, examination can be directed towards those areas that most significantly affect the overall time. In reducing the times of the initially critical activities, new critical paths may be created which, in turn, must be scrutinized.

The questioning method

In order that the examinations of the various activities will be consistently useful, it is desirable to employ the well-tried work study technique of systematic questioning. In this, a number of questions are set up, and these questions asked of every activity. By asking the same questions in this apparently rigid way it is possible to ensure that a thorough examination is made of *all* alternatives. For a full discussion of this method, reference should be made to any one of the many textbooks on work study. The following should be considered only as an introduction to the method.

Activities can be considered to be of two kinds:

(1) 'DO' activities, where time is consumed in a task which, in itself, advances the project.
(2) 'ANCILLARY' activities, where time is consumed in tasks which support 'DO' activities.

For example, if a project involves the making of a component, the act of making is a 'DO' activity, and the acts involved in setting up and breaking down the plant to carry out the making are 'ANCILLARY' activities. Clearly, it is the 'DO' activities that should be examined first, since if they can be reduced or eliminated, the 'ANCILLARY' activities may either vanish or be reduced. (The author recalls a project that involved an activity 'assemble refrigerated tank,' along with its associated activities of 'place orders,' 'obtain material,' 'test lagging' and so on. Discussion had centred on the problems of reducing purchasing time, until the 'DO' activity—'assemble refrigerated tank'—was examined, when it was discovered that in fact this was an entirely unnecessary activity, and with its elimination, the ANCILLARY activities disappeared.)

Once these 'DO' items on the critical path have been identified, they can be tested against a series of questions, which are dealt with more fully in R. M. Currie's *Work Study*.

(1) *Purpose*: What is being done?
Why is it being done?
What else could be done?
What should be done?

(2) *Place*: Where is it being done?
Why there?
Where else could it be done?
Where should it be done?

(3) *Sequence*: When is it done?
Why then?
When else could it be done?
When should it be done?

(4) *Person*: Who does it?
Why that person?
Who else might do it?
Who should do it?

(5) *Means*: How is it done?
Why that way?
How else can it be done?
How should it be done?

Should *these* questions, having been applied to the critical activities, not produce the desired result, one other question may be asked, although it must be stressed that it is a dangerous one to ask, namely:

(6) Can an activity be reduced by increasing the *risk*?

For example, it may be that the initial network has in it a 'testing', 'checking' or 'proving' activity. Such activities can often be reduced, but with an increase in the risk of failure; thus, after drawings are completed, checking is often carried out, and if this is thorough then the checking time can be great—a substantial part of the initial drawing time. If this checking time is reduced, there is a greater chance that errors will slip through, with all the consequent undesirable results. To reduce this time, therefore, will increase the risk, and this decision must be squarely put to management for acceptance or rejection.

Reduction involving transference of resources

The non-critical activities in a network can sometimes be used to obtain resources that can be applied to critical activities to reduce their durations. This is sometimes known as 'trading off' resources.

Reduction involving increased cost

All other methods having failed, a reduction in time may have to be obtained at an increased cost, usually by increasing the resources that are employed. If the costs to reduce times are known, then a table can be set up showing the relative costs for the reduction in time of each activity by the same amount. The cost incurred in reducing duration time by unit time may be defined as the 'cost-slope' thus, for activity *B*:

Normal duration time of 20 weeks costs £200
Reduced duration time of 19 weeks costs £220

hence, the cost-slope = £20 per week.

For the sample network already considered, the table of costs slope might be:

Activity	*Duration*	*Total float*	*Cost-slope (£/week)*
A	16	8	30
B	20	0	20
C	30	21	10
J	15	8	60
D	15	0	45
E	10	9	120
G	3	1	10
H	16	0	15
K	12	1	95

Clearly, of the critical activities, activity *H* has the smallest cost-slope, and it is desirable to investigate the practicability of reducing it first. These investigations may show not only that it can be reduced by 1 week, at an increased cost of £15, but further reductions are also readily obtainable. Inspection of total float, however, shows that two activities (*G* and *K*) will become critical if activity *H* is reduced by 1 week and, in fact, two critical paths:

B–D–H and *B–D–G–K*

will be formed. Thus, in order to reduce total project time (*TPT*), *either* the common part of these two paths (i.e. *B–D*) must be reduced *or* the two branches (*H* and *G–K*) must be reduced simultaneously. The least expense is incurred when reducing the two branches by shrinking activity *G* (cost-slope £10/week) and activity *H* (cost-slope £15/week). This will produce an effective cost-slope of £25/week, which is greater than the cost-slope of activity *C* (£20/week), so that it would probably be desirable to investigate the reduction of activity *B* first.

Just as it is possible to reduce duration times by increasing costs, so may it be possible to reduce costs by increasing times. For example, activity *C* as planned has a duration time of 30 weeks and a total float of 21 weeks. Examination of activity *C* may show that its duration time could be increased to, say, 40 weeks, while at the same time reducing the cost. Such a reduction in cost would not increase the overall project time, but the savings might help to offset the increased costs of shrinking other duration times.

By means of this sort of approach, overall project times can be reduced and total costs minimized. In a simple network as discussed here, no particular difficulties will arise, but in larger networks the number of alternatives will be very great indeed, and it may therefore be necessary to employ a computer for this work.

The dangers of the cost-slope concept

The concept of 'cost-slope' is appealing in its simplicity. However, it must be pointed out that:

(1) It is frequently extremely difficult to obtain reliable figures for the changes in cost resulting from changes in duration time. These difficulties are so great that in practice the cost-slope technique is unusable. The author has never found any example of the useful adoption of the technique.

(2) The relationship between cost and time is not a simple one. Multiplying labour time by wage cost is obviously inaccurate and, moreover, to 'extend' the resultant labour cost by a constant overhead factor can be equally misleading, since the reduction in time may be obtained, for example, by hiring special plant that has a non-linear hiring rate.

These difficulties make it dangerous to try to deduce general *time-cost* curves, or, to put it another way, to assume that cost slopes are constant. For *short* time intervals this assumption can be reasonable, but it is desirable to examine it very closely. All this, of course, is true whether PNT is used or not, but employing PNT has the great advantage that investigations can be directed to the critical activities.

The relationship between the time and labour employed

Duration times cannot be reduced indefinitely by increasing resources. For example, in digging a hole it may well be that two men can carry out the work in less than half the time that one man can carry it out, since work can be efficiently divided. However, three men may not show the same reduction in performance time, and a fourth man may well slow up the work since his physical presence may impede the other workers (Fig. 13.1). There is thus a minimum time below which it is not possible to reduce the duration time of an activity.

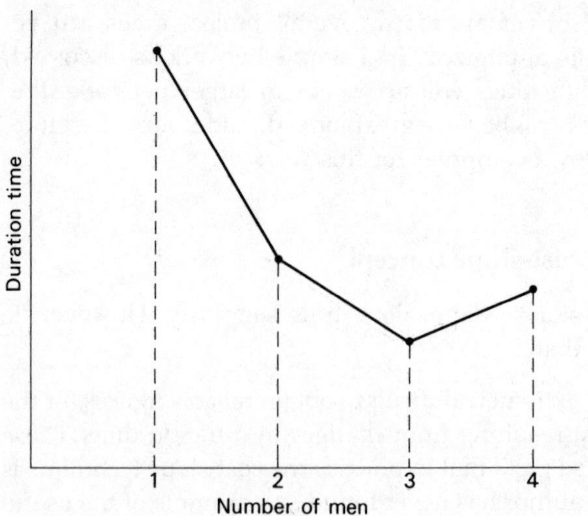

Fig. 13.1

The final network

The final network, after a reduction process has been carried out, may well be considerably different from the initial network. The logic may have changed, duration times altered and new critical paths created. Illogicalities may have been introduced, and it is worth retesting the network by checking every activity once more against the two questions:

(1) What had to be done before this?
(2) What can be done now?

Check especially any 'cross-road' situations to ensure that *all* emerging activities do depend upon *all* entering activities.

Questions to be asked when reducing the project time

Purpose?
Place?
Sequence?
Person?
Means?
Risk?
'Trade-off'?
Cost?

14 Resource allocation: I Basic considerations

In all the discussions so far, it has been assumed that *time* is the most important factor, and in very many cases this is indeed so. However, it may well be that not only is time important, but the resources employed are equally, or more, important; for example, the number and skill of the people employed, or the special equipment used may impose severe restrictions.

This problem is one that has been well known to the production control engineer since any form of activity has been undertaken. The name given by production controllers to this aspect of their work is *loading*; regrettably, new names have been devised by some of the earlier project network techniques (PNT) workers, among them *man-power smoothing* and *resource allocation*. In the present text resource allocation will be used.

Loading is defined as 'the assignment of work to an operator, machine or department,' and is a most important feature of producing a time-table. When too much work is required of a work source, the work source is said to be *overloaded* while if too little is needed it is said to be *underloaded*. Ideally, the work required should be exactly equivalent to the work available, when the work source is *fully loaded*. This is an ideal situation which is seldom, if ever, encountered except in large-scale flow production, where it is possible to adjust supply and demand in order to reach some form of parity. In the type of work for which PNT is most useful, it is frequently impossible to adjust both supply and demand, and some form of compromise is essential. This usually takes the form of *underloading* since this, at least, produces an acceptable result with respect to time; that is, the promised delivery date can be met. Deliberate *overloading* is foolhardiness to the point of sheer irresponsibility. Projects whose starting and finishing dates are fixed are said to be 'time-limited', while those where available resources are limited are 'resource-limited'. Projects that at both time- and resources limited are likely to be unachievable, and managerial action will be necessary to release one constraint.

Work required

Resource allocation requires, as a point of departure, a statement of the work required. This can only be given in terms of man- (or machine-) hours; it should not be given in any other units *unless* those units are readily and acceptably capable of being translated into resource-hours. For example, it may be that to dig a hole 4 ft × 4 ft × 6 ft in a particular location would take one man 12 hours. The work content should then be specified as 12 man-hours of work, *not* as 96 cubic feet of digging, unless it is well established that:

12 man-hours of digging = 96 cubic feet

or

1 man-hour of digging = 8 cubic feet

Similarly, if a designer is committed to design three transformers, it is meaningless to state that his work load is three transformers unless it has previously been established that one transformer is equivalent to so many hours' work.

The work required—that is, in more usual terms the work content—is thus specified in resource-time. More than this, it is necessary also to specify the method employed; the work content of the above 4 ft × 4 ft × 6 ft hole is said to be 12 man-hours when one man is digging. If the method of working is changed, for example, by adding another man to clear away the loose earth, the total time taken may be reduced to 5 hours—that is, the work content is then 10 man-hours.

The work required is initially specified by reference to the *usual* way of carrying out the activity, with the *usual* methods and the *usual* resources, working at the *usual* rates. Initially, no cognizance is taken of the apparent need of other activities that may require the same resources—'infinite capacity' is assumed at the outset. If necessary, limitations on resources can be considered later. It is useful to establish a library (databank) of information concerning tasks likely to be repeated, their durations, methods and resources, both planned and achieved.

Work available—capacity

To complete the task of loading, it is necessary to know the amount of work *available*—that is, the *capacity* available. This, too, must be specified in resource time, but in order that the completed schedule will not be fictional it is essential that the capacity should be strictly realistic. Thus, if 100 men are employed, it is unwise to assume that the available capacity each week is 100 man-weeks. In calculating available capacity it is necessary to know:

(1) The usual efficiency of working.
(2) The anticipated sickness or absenteeism.
(3) Existing commitments.
(4) Anticipated maintenance work which is to be done.
(5) Holidays.
(6) Any other limitations on working, for example:
    (a) confined space.
    (b) limited machine capabilities.
(7) The possibility of extending capacity by, say, overtime or sub-contracting. Such extension inevitably carries a cost and this too must be known.

In practice, it may well be that some of these factors may be ignored, but it is probably wise to consider each of them as a matter of routine; the number of projects that have not been completed on time because the presence of the summer holidays was forgotten is not on record, but it is uncomfortably high.

Calculation of load—resource aggregation

The network will specify, on each activity, not only the desired time, but also the resources required. The total resources for each time interval are then calculated by moving through the network, one time interval at a time, and adding up the resources of each type during each time interval. This process—resources aggregation—is initially carried out with each activity starting *as early as possible*. These aggregated loads are then compared with the available capacities. When overloads occur, attempts are made to reduce them by 'sliding' activities along within their floats. If the time for the project is fixed then the amount of float that is possible is also fixed and the project is 'time-limited'. It may not, with this limitation, therefore, be possible to match load and capacity. On the other hand, if time is not fixed but capacities are fixed, then it may well be that the act of matching load and capacity is to increase the total project time (*TPT*) beyond that which was initially calculated.

Resource allocation

Once aggregation has taken place, allocation—the distribution of resources across the project—will almost inevitably follow. The author's experience is that in any sizeable project it is unusual to find that the resource requirement as displayed by the initial resource aggregation is acceptable. In carrying out allocation, at least three problems arise:

1. Imperfection of the data

The peculiar juggling act of resource allocation requires knowledge of:

(*a*) Activity logic and duration times.
(*b*) Resources required.
(*c*) Resources available.
(*d*) Priorities applied to tasks and resources.

Unfortunately, all these are liable to be known imperfectly, and to change as the project progresses. To try to make firm decisions 'cast in concrete' is foolish. What is required is the ability to make resource allocation decisions *fast*, so that as changes take place and errors come to light they can be accommodated in 'real-time'.

2. The problem of optimization—the objective function

It is often loosely stated that an 'optimum' allocation is calculated. The problem here is that it is frequently difficult, if not impossible, to decide which feature should be optimized, that is the *objective function*. Thus, if to complete a task it is necessary to utilize men and machines, and also to purchase material, there are at least four different aspects to consider. Should the programme be constructed to:

(*a*) Make maximum use of labour?
(*b*) Make maximum use of machines?
(*c*) Hold purchased materials for as short a time as possible?
(*d*) Increase customer goodwill by reducing the overall time as much as possible?

Regrettably, these requirements are often in conflict, and, since it is difficult to assign objective values to any of these, and even more difficult to foresee the inter-relationships between the various factors, the decision must often be made on arbitrary grounds. Once a decision has been made, it may be necessary to change the decision as the project progresses.

The essential feature, of course, is that the problem must be considered and a decision must be made and continually reviewed. It is sometimes convenient to rank the various factors, and then attempt to optimize 'in sequence', i.e. try first to optimize factor 1, then factor 2 and so on, not allowing any succeeding factor to affect any of its predecessors.

3. The problem of alternatives

The problem here is one that arises from the interaction of one job upon another. For example, if in a department there are three operations, *A*, *B* and *C*, to be carried out, these operations being independent of each other, then it may be seen that there are six possible sequences in which work can proceed:

(1) Do *A* first, followed by *B*, then by *C*.
(2) Do *A* first, followed by *C*, then by *B*.
(3) Do *B* first, followed by *C*, then by *A*.
(4) Do *B* first, followed by *A*, then by *C*.
(5) Do *C* first, followed by *B*, then by *A*.
(6) Do *C* first, followed by *A*, then by *B*.

Similarly, if there are four operations, there are 24 possible sequences, and if there are *N* operations, there are *N*! possible sequences, and *N*! increases very rapidly.

Of the possible sequences one (or more) may produce the solution that is 'optimum' according to the rules laid down, but to *know* this by calculating all the possible sequences is virtually impossible. The difficulty is seen clearly if *N*! is set out for the first few natural numbers:

N	*N*!
1	1
2	2
3	6
4	24
5	120
6	720
7	5,040
8	40,320
9	362,880
10	3,628,800

and to have 10 tasks that should be carried out in one department/section during any one planning period is extremely modest. Further, only one department has been considered. If there are a number of departments, then the number of possible sequences becomes astronomical.

PNT is of some value here, since it is possible to concentrate upon the critical activities, and this may reduce the choice very considerably. Even so, it may not be possible to examine all solutions, and it is often better to take an *acceptable* solution—that is, one which is workable and does not offend any optimization rules too much—than to spend a great deal of effort in trying to find the probably unrecognizable optimum solution.

While it is probably impossible to lay down any fundamental laws on loading, it would seem reasonable to:

(1) Define the allocation rules, and then
(2) Adopt an 'acceptable' solution,
(3) Using a method of calculation that gives a *fast* result,

in order to reduce the loading problem to easily manageable proportions. It should be realized that, except in limited cases, even a computer cannot calculate all the possible combinations. Two general approaches are used:

(1) *The decision rule approach*: Here, a set of rules, which may have no other justification than that they 'feel right', is applied to the network. One such set of rules is illustrated in the next chapter.
(2) *The optimum seeking approach*: Here, a mathematical algorithm designed to produce an optimum is used. There are several possible methods available—one is illustrated in Chapter 16.

A limited case

The general problem of resource allocation is, as has been seen, an extremely complex one, and it will be illustrated here by a very simple example.

The load as a histogram

It is convenient to represent the load as a histogram—that is, a vertical bar graph, the length of the bar being proportional to the load. For example, if the weekly load in a department that has a capacity of 10 man-weeks is:

Week No.	*Load (man-weeks)*
1	6
2	7
3	8
4	10
5	12
6	6

this would be represented by Fig. 14.1, which shows very clearly that the department is underloaded in weeks 1, 2, 3 and 6, fully loaded in

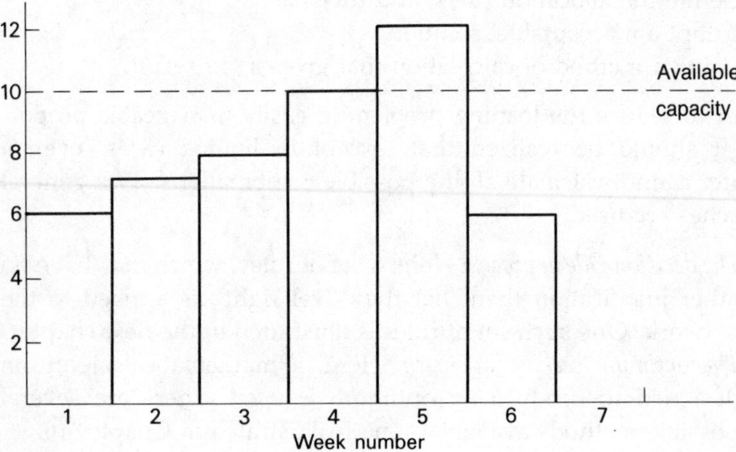

Fig. 14.1

week 4 and overloaded in week 5. (*Note*: some workers in this field refer to the available capacity as the manpower ceiling, so that week 5 exceeds the manpower ceiling.) Although this representation gives no more information than the corresponding sets of figures, it has the usual virtue of a graphical representation, namely great vividness, and in practice it is found that it is almost invariably easier to work in histograms than in numbers, even though the histograms have been derived from the numbers. With experience, a great facility is obtained in viewing a histogram and assessing whether a 'peak' (i.e. an overload) can be toppled into a 'valley' (i.e. an underload) in order to 'smooth out' the loading.

Drawing the histogram by hand

The simplest way of drawing a histogram is probably found by drawing the appropriate Gantt chart and, by running down each time division, adding up the usage of the various resources. For example, consider the network that has been so frequently discussed (Fig. 6.3, p. 53 and Fig. 9.1, p. 88) and assume, for the purposes of illustration, that the only resource used is *men*, and that each activity requires manpower as follows:

Activity	*Duration*	*Men*
A	16	2
B	20	6
C	30	4
J	15	3
D	15	2
E	10	5
G	3	2
H	16	4
K	12	4

Re-draw the Gantt chart, as described in an earlier chapter, inserting man-requirement on each activity bar within a circle. Then, by running down each week, it is possible to add up the manpower requirements very simply, and these can then be plotted on a histogram which, for the sake of convenience, will be shown beneath the Gantt chart (Fig. 14.2).

In practice, it is possible, by the exercise of a little common sense, to reduce the number of additions; for example, it is clear that the loading for weeks 1–16 is the same, so that one addition $(2+6+4)$ suffices.

If the capacity is also inserted on the histogram, the labour situation is very clearly shown. Assume that the available capacity is 10 man-weeks, and that all men are interchangeable. The dotted line shows this capacity and the over- and underloading. A quick check can show at this stage if the available capacity is adequate. The resource time commitments are calculated and totalled:

Activity	Resource time
A	32
B	120
C	120
J	45
D	30
E	50
G	6
H	64
K	48
Total	515

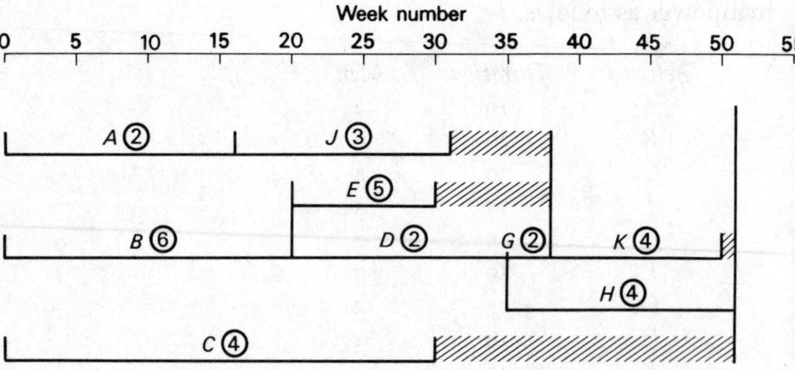

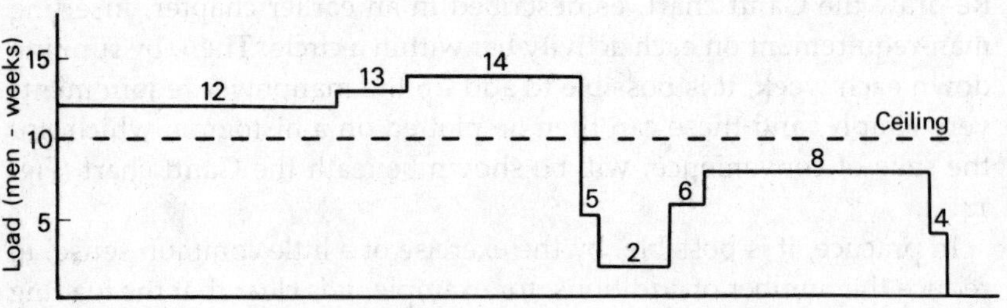

Fig. 14.2

The *TPT* is 51 weeks, hence the minimum possible resource demand is 515 ÷ 51 men = 10.1 men. Two choices are now possible: set a new ceiling of, say, 11 men, or use the existing ceiling and then see where an increase of resources is necessary. Since the 'theoretical' demand is so near the ceiling, the second course is followed.

Smoothing the load

The situation revealed by the histogram is one that is completely unacceptable. For 30 weeks the load exceeds the capacity, which can have only one result, namely that activities will take longer than planned, and the overall project time will increase. For 21 weeks the capacity exceeds the loading, and this will mean that men are idle. Clearly it is desirable to try to shift some of the earlier overload into the later underload. If this could be completely done, then the load would be 'smoothed'.

This problem is an extremely familiar one to all those who have been in charge of the organization of the disposition of labour. Virtually intractable, PNT does assist by providing guidelines along which to work. Of the various activities some are fixed in time (that is, some are critical), while others can move (that is, they possess float), and, if an increase in *TPT* is to be avoided, smoothing must take place in the 'floating' portion of the load. Furthermore, significant changes can only be made where float is substantial. Thus, activity *G* has only one week free float; its overall effect is therefore small.

Activity *C* possesses the greatest float, and it should therefore be examined first. It will be seen that its duration is so great that, while floating it as much as possible will reduce the load at the beginning of the period, it will not seriously reduce the 'lump' between weeks 20 and 30. The next activity, in order of magnitude of float, is activity *E*. If this is moved as much as possible, activity *K* will advance by 1 week and activity *E* will extend from week 29 to week 39. This will then give a Gantt chart and histogram as shown in Fig. 14.3.

The overload has been completely removed during weeks 20–29, and a small overload has been introduced during weeks 35–38. All float has been removed from activities *E* and *K*. The remaining activity with any substantial float is activity *J*. Shifting this forward by 4 weeks would reduce the 'lump' during weeks 16–20 and fill the 'trough' during weeks 31–35. The chart and histogram would then look as shown in Fig. 14.4.

Moving any other activities—and there are only activities *A*, *C* and *G* that *can* be moved—would produce no significant change in the load. Hence, the arrangement in Fig. 14.4 is that which gives least overload and, using this as a criterion, is the 'best' arrangement.

(*Note*: In practice, the virtue of any particular arrangement can only be judged within the context of the local circumstances.)

The above discussion is offered as an illustration of the kind of

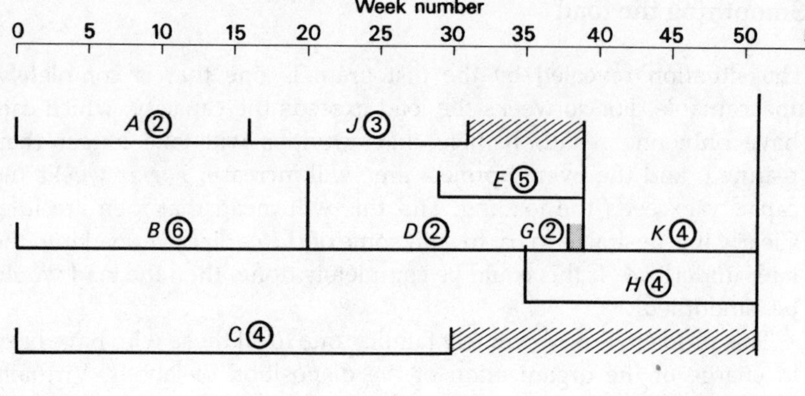

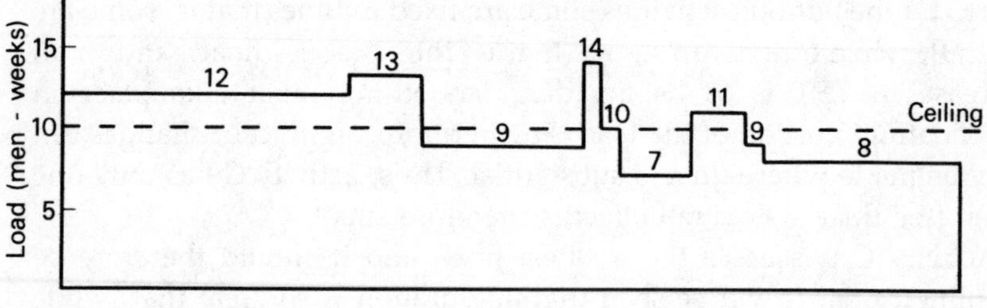

Fig. 14.3

thinking that lies behind a loading task. The answer arrived at is not ideal, but answers seldom are in practice. The result, however, does give a sound basis for further consideration by which the problem can only be resolved managerially. For example:

(1) The 'spike' in week 29 can be removed if activity *E* is advanced by one week, but this will move the critical path forward by 1 week and hence the overall project time will increase from 51 to 52 weeks. Is this desirable/acceptable?

(2) The overload during weeks 1–16 can be removed by splitting activity C into two parts, the first 20 weeks long, the second 10 weeks long, and performing the second part during weeks 42–51. Is this desirable/acceptable/possible?

PNT does not solve the resource allocation problem, but it does provide a method for systematically examining the possibilities. If there are more resources than one, then the examination becomes correspondingly more difficult.

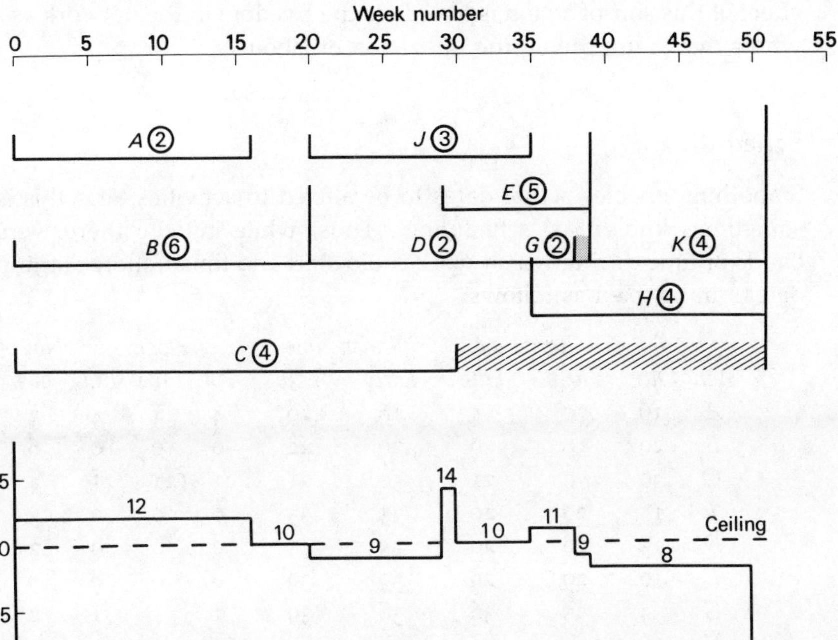

Fig. 14.4

The effect of smoothing

What is the effect of smoothing on the project as a whole? To examine this, assume that the last situation above (Fig. 14.4) is taken to be the acceptable one. This requires that activity *E* should not start until week 29, and that it must finish on week 39. Thus, the earliest and latest start-time (*EST* and *LST*) for activity *E* is week 29, and the earliest and latest finish time (*EFT* and *LFT*) is week 39 and, since the duration time is 10 weeks, all float has disappeared. Effectively, another activity, 'wait for availability of labour for activity.' has been inserted, and the relevant part of the arrow diagram has changed to include a new activity, *W*, 'waiting for labour', duration 9 days, which precedes activity *E* and succeeds activity *B* to create a second critical path:

B–W–E–K

in parallel with the first. Similarly, fixing the starting time of activity *J* at not earlier than week 20 has removed the float in the activity. The effect of this sort of action is to reduce the freedom in the network as a whole, while improving the utilization of labour.

Scheduling

Smoothing enables actual dates to be affixed to activities, and this is sometimes known as 'scheduling'. Thus, while initially there were bands of time during which work could start and finish, more starting dates can be fixed as follows:

		Start time		*Finish time*		*Float*			*No. of*
Act.	*Dur.*	*Early*	*Late*	*Early*	*Late*	*Tot.*	*Free*	*Ind.*	*men*
A	16	0	4	16	20	4	4	4	2
B	20	0	0	20	22	0	0	0	6
C	30	0	21	30	51	21	21	21	4
J	15	20	20	35	35	0	0	0	3
D	15	20	20	35	35	0	0	0	2
E	10	29	29	39	39	0	0	0	5
G	3	35	36	38	39	1	1	1	2
H	16	35	35	51	51	0	0	0	4
K	12	39	39	51	51	0	0	0	4

On comparing this to the original network analysis it will be seen that much of the original element of float has disappeared.

Two types of scheduling: serial and parallel

If a project requires more than one type of resource then, in general, scheduling can proceed in one of two ways. Both involve ranking the resources in order of importance or significance to the project. For simplicity, assume that the resources (types *A*, *B*, . . .) have ranks corresponding to their alphabetic sequence, that is, that type *A* is the most important and has therefore rank 1, type *B* is the next most important, rank 2, and so on.

Serial scheduling

Here the rank 1 (*A*) resource is scheduled throughout the life of the project. Rank 2 (*B*) is then scheduled throughout the project, then rank 3 (C) and so on until all resources are allocated.

Parallel scheduling

Here the project is divided into convenient time periods and the resources scheduled in order of rank, throughout the period. At the conclusion of the period a new rank *may* be assigned to each resource if this seems appropriate, and again each resource is scheduled.

Considerable debate has taken place concerning the relative merit of each system. Given the usual imperfection of data (time, resource requirements, ceilings), the search for the 'best' schedule is probably a pointless task. In practice, it is necessary to respond to changes *quickly* and, therefore, whichever method gives the quickest feasible schedule should be adopted. It is rare to schedule a complete project except at the 'feasibility' or 'estimate' stages where fairly generous approximations are made. Most computer programs have a facility for changing resource rankings as often as required.

Sharing resources between projects

Float, as has been stated, is equivalent to an under-utilization of resources. To increase utilization it is common practice to employ the under-utilized resources on another project. For example, it might be possible to transfer the four men needed for activity *C* to another project for some or all of the 21 days' float. This will increase the utilization of the four men and effectively smooth two projects as a whole. It will, however, have the same effect as single-project smoothing, namely that it will introduce another activity, 'waiting for labour from project . . .', and in turn this will reduce flexibility.

As with loading, the sharing of resources is a difficult task, and it can only be systematically carried out in very restricted cases. In general, the loading and/or sharing of labour is an empirical process, where a workable answer must be accepted, even though it cannot be demonstrated to be the optimum answer. It is in this area that the computer can probably be of maximum assistance, since the computer can work out a large number of solutions very rapidly. It is also this area (which is very similar to areas in other, closely allied, fields) which is receiving enormous attention from research workers, and it may well be that before long this whole problem can be dealt with simply and systematically, although there will always be the need for management to decide on the criteria against which solutions will be judged—a difficult task.

15 Resource allocation: II A decision rule approach

Despite the difficulties discussed in the previous chapter, the need to produce a solution—albeit an imperfect one—to the resource problem is often encountered. It is useful here to remember Bowman's comment (see *Industrial Scheduling* edited by Muth & Thompson, Prentice-Hall Ltd.) that almost any set of sensible decision rules, if applied consistently, is likely to produce a 'better' schedule than an *ad hoc* series of decisions applied unsystematically. The present chapter will illustrate one such set of decision rules.

Decision rules

When two or more activities require resources in excess of those available, a *conflict* arises that has to be resolved. For example, in Fig. 15.1, activities *P* and *Q* both precede *R* and activities *P* and *Q* require 3 and 4 units of a resource respectively. Only 5 units are available, and during the period day 20–day 30 a total of 7 units are needed. Conflict therefore exists between the two activities and this can be resolved by 'awarding' the available resources either to activity *P*, in which case activity *R* must be 'slid' to the right, or to activity *R*, in which case it is the other activity that is moved.

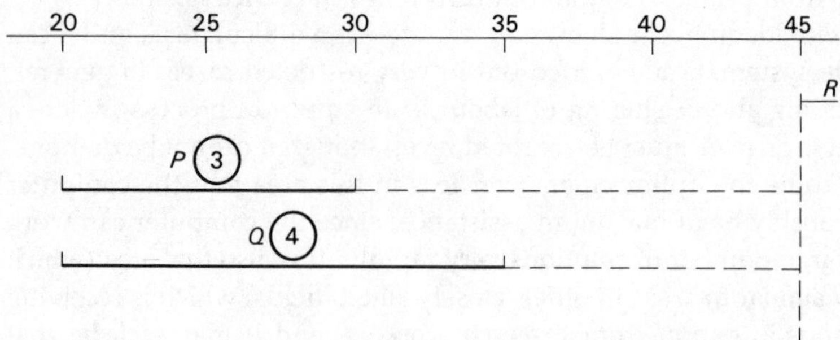

Fig. 15.1

With a multi-activity network it is not possible to test all feasible alternatives, and some routine must be set up to arrive at a decision.

This involves the use of rules, known as *priority* or *decision* rules, which can rarely, if ever, be justified on any logical grounds, and the results they provide cannot be assumed to be optimal in any way. Many rules can be devised, but for the purposes of illustration a set of rules derived from the work of Dr Martino will be used here. While these have considerable intuitive appeal, no claim for superiority is made for them.

Gordon, in a substantial study of heuristic resource allocation techniques, reports that 'the use of latest start as a priority ranking heuristic gives results that are as good as, or better than, those that can be obtained by any other heuristic rule'. He also suggests that if activities can be 'split'—that is, once started can be stopped and re-started later—'the use of a parallel system . . . is verging on an absolute necessity . . . but when an activity once scheduled must be completed without interruption . . . the use of a serial system has obvious advantages in terms of computer time required. . . .' (See Gordon, J. H. 'Project Network Categorization in an Evaluation of Heuristic Resource Allocation Techniques', PhD thesis, May 1975, University of Birmingham.)

Most computer programs for resource allocation incorporate sets of rules, and often these can be varied by the user. The reader is advised to write his or her own resource allocation program only *in the most exceptional and extraordinary circumstances*. These programs are tedious and time-consuming to write, and many tested programs are already on the market.

Manual resource allocation

A manual application of a set of decision rules is illustrated here; in cases where it is too difficult or too unwieldy to carry out manipulations by hand, a computer can be used. Considerable insight into computer methods can be obtained by following through the manual technique below.

Step 1. Determine the resources required

For the standard network shown below, assume that the resources required are:

Activity	*Duration (weeks)*	*Work content (resource/weeks)*	*Type*	*Resources required*
A	16	32	*w*	2
	16	48	*x*	3
	16	32	*y*	2
B	20	20	*x*	1
	20	60	*y*	3
C	30	60	*w*	2
	30	60	*x*	2
	30	90	*y*	3
	30	90	*z*	3
J	15	45	*w*	3
	15	15	*x*	1
	15	60	*z*	4
D	15	15	*z*	1
E	10	30	*w*	3
	10	40	*y*	4
	10	40	*z*	4
G	3	12	*x*	4
H	16	32	*w*	2
	16	64	*y*	4
	16	48	*z*	3
K	12	48	*w*	4
	12	48	*z*	4

Note: 'Duration,' 'work content' and 'resources required' are all interdependent, any one being deducible from the other two. Circumstances can arise when a different duration must be used in order to change the resource requirement.

Step 2. Determine the resource ceiling

In some situations the resource ceiling is known ('the maximum number of type *x* resources that can be made available is . . .') but in other circumstances the requirement is to try to use the minimum possible resources. In this case, proceed as follows:

(*a*) Carry out a forward and backward pass to determine the total project time (*TPT*). If this is acceptable, then loading may proceed. If it is unacceptable, then it will be necessary to reduce the *TPT* by methods similar to those suggested in Chapter 13. *Do not, at this stage, attempt to decide on the disposition of limited resources.* In the standard network *TPT* = 51 weeks.

(*b*) Calculate the total work content for each resource type:

Activity	*Resource type*			
	w	*x*	*y*	*z*
A	32	48	32	0
B	0	20	60	0
C	60	60	90	90
J	45	15	0	60
D	0	0	0	15
E	30	0	40	40
G	0	12	0	0
H	32	0	64	48
K	48	0	0	48
	247	155	286	301

(*c*) Divide the total work content by the *TPT* (here, 51 weeks).

Resource			
w	*x*	*y*	*z*
4.8	3.1	5.6	5.9

(*d*) Take as a resource ceiling the next whole number above the figure obtained in (*c*) unless that number be itself a whole number, when this is taken as the resource ceiling:

Ceiling for resource type			
w	*x*	*y*	*z*
5	4	6	6

Step 3. Prepare a bar chart

Translate the network into a bar chart with all activities starting as early as possible. The Gantt chart shown in Fig. 15.2 is convenient for this purpose, but any type of chart may be used. Inscribe on the bar chart a statement of the resources required for each activity.

Step 4. Determine total resource requirements (aggregate resources)

For each period of time, sum up the resources required and the resource ceilings. While a histogram is vivid, for a number of resources numerical statements are more convenient. This is the process defined as resource aggregation in BS 4335: 1972. 'The summation of the requirements of each resource for each time period.'

Note: It is helpful at this stage to rank the resources in importance and to set them down in decreasing order. In Fig. 15.2, it is assumed that resource w is of greater importance than resource x which is more important than y which in turn is more important than z. In practice, it is often found that this 'priority ruling' changes through the life of the project as both internal and external circumstances change. Again, most computer programs can easily accommodate such changes.

Step 5. Establish decision rules

Where there is a conflict for resources, award resources according to the following rules devised by Martino:

First priority	in order of float
Second priority	in order of work content
Third priority	in order of size of resource
Fourth priority	in order of priority of resource
Fifth priority	in order of latest finish
Sixth priority	in order of j number

Sets of rules such as these can often be adopted as standard rules within an organization.

Step 6. Utilize float to cause demand to coincide with capacity

(*a*) Starting at time 0, run a cursor along the date scale until an overload situation is observed. With the present example this in fact occurs at week 1 where 6 units of x are required (ceiling = 4) and 8 units of y are required (ceiling = 6). To relieve these overloads it would be possible to move either of activities A or C. Priority is determined (step 5) in order of float: activity A has less float than activity C. Hence, resources are 'awarded' to activity A, and activity

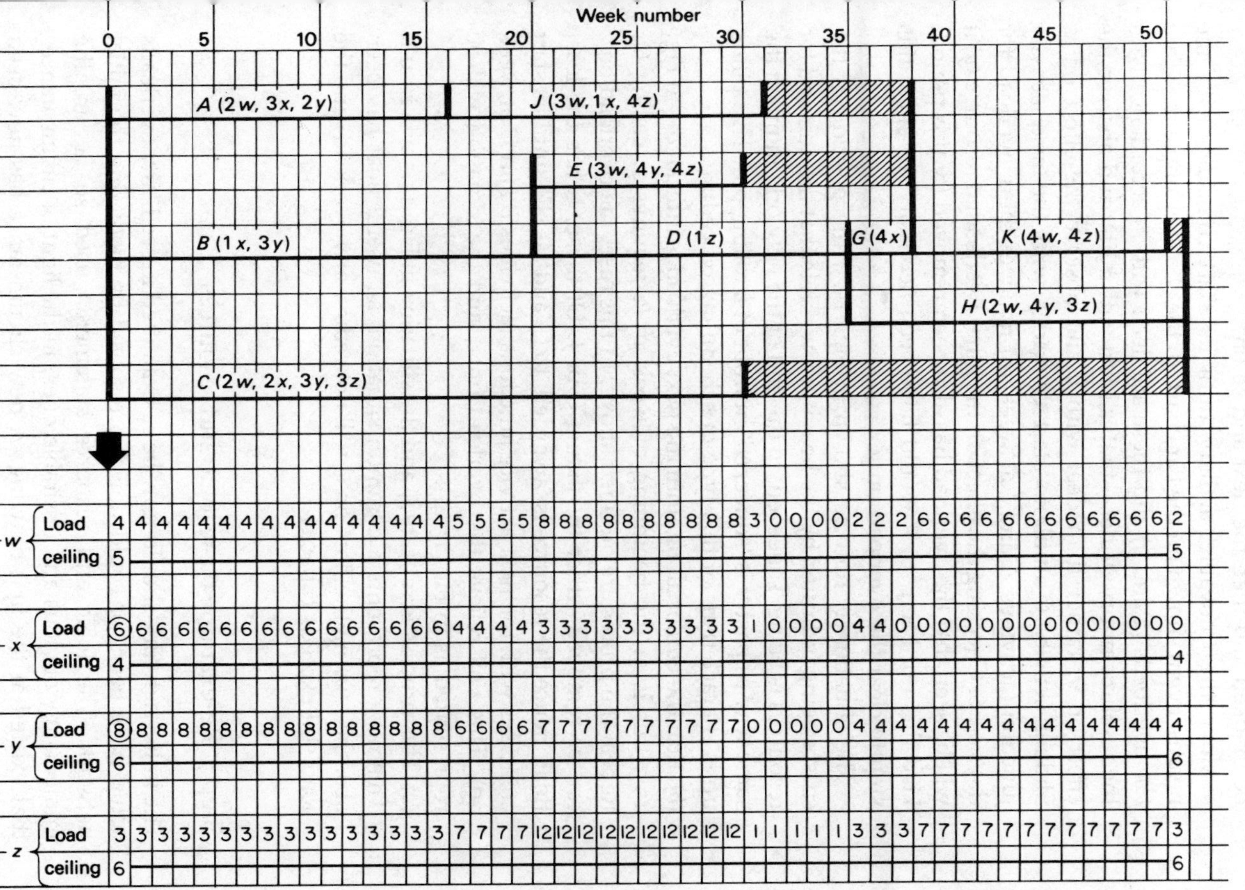

Fig. 15.2 Initial resource aggregation

C is moved. If *A* and *C* had had equal float, then the next rule would be invoked and then the next and so on.

Move the 'movable' activity until its start coincides with the nearest event shown on the bar chart. In the above example there are *two* overloads (*x* and *y*) and these are dealt with in accordance with the resource priority rules—that is, deal with *x* first and observe the effect on *y*, rather than deal with *y* and observe the effect on *x*. Sliding activity *C* until its start coincides with the next activity junction (activities *A* and *J* at week 16) removes the *x* overload for days 0–16. Checking the effect on the other overloaded resource (*y*) it will be seen that the overload has also been removed for this period. Had this not been so, it would have been necessary to repeat this step to clear the overload on *y*.

The situation is now as shown in Fig. 15.3, the position of the cursor being indicated by the heavy arrow. Note that the demand for resources is being 'squeezed' to the end of the network, in much the same way as pastry is 'squeezed' in front of a rolling pin. This piling up is characteristic of this type of scheduling procedure.

(*b*) Run the cursor along until the next overload is indicated—in Fig. 15.3 this is at the beginning of week 17 (*z* requires 7, ceiling 6). Repeat the procedure outlined above. In this case, the overload on *z* can be removed either by sliding *C* or *J*. Of these, *C* has the least float, hence the resource is 'awarded' to *C*, and *J* is slid until its start coincides with the next activity junctions (activities *B* and *D* at week 20). This will relieve the overload on *z*, and the situation will be represented as in Fig. 15.4.

(*c*) This process is repeated, and the situation in Fig. 15.5 will result. This now represents a common situation: all useful float has been absorbed (using the float on *C* will not reduce any overload) and the load still exceeds capacity.

Step 7. Re-examine basic logic and resources

The initial network should now be re-examined to see if any overloads could have been, or could now be, resolved by modifying the initial network. The logic and resource statements used so far are the immediately obvious and desirable ones, in the light of circumstances that existed at the first drawing. A new circumstance has now been created, and the original logic and resources used produce an unacceptable situation. Thus, as when any budget is shown to be

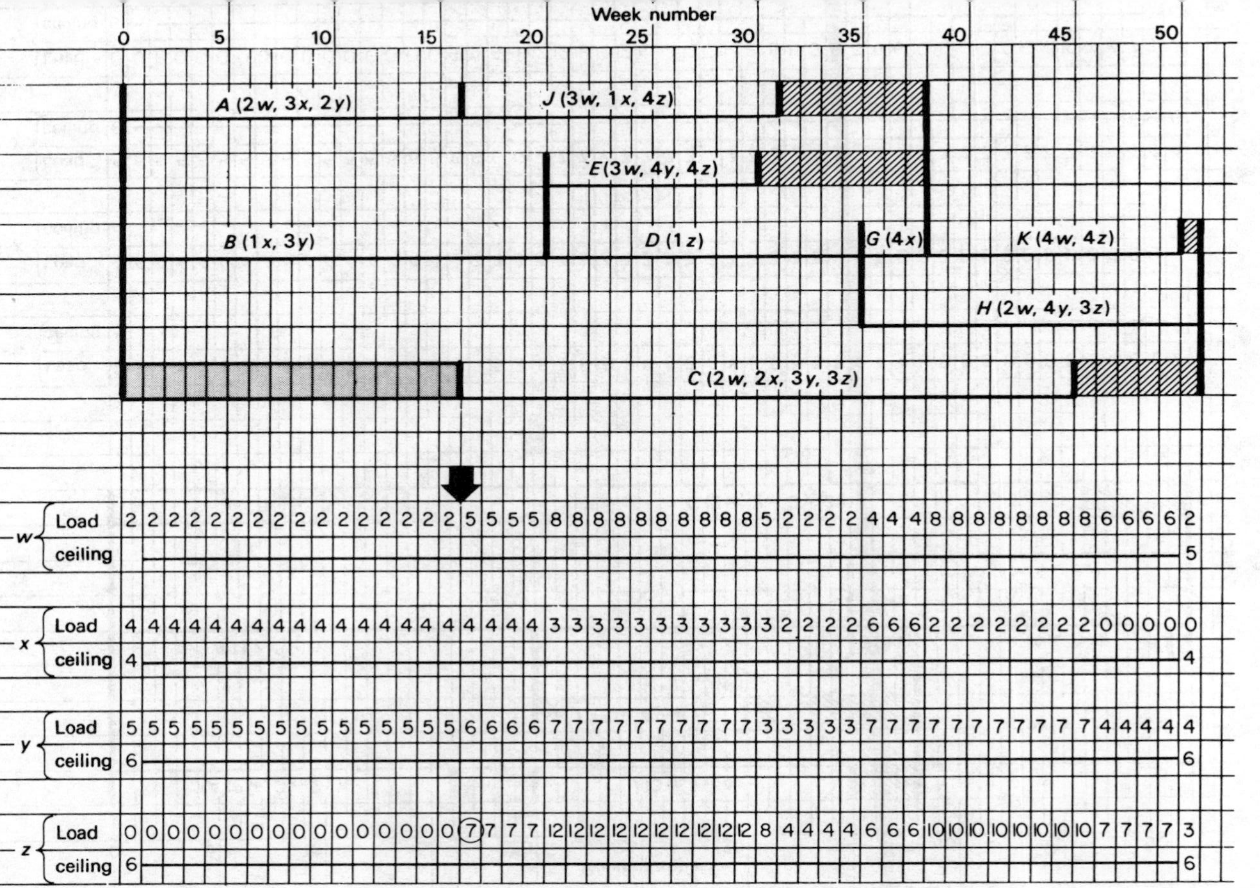

Week number
0
5
10
15
20
25
30
35
40
45
50
A (2w, 3x, 2y)
J (3w, 1x, 4z)
E (3w, 4y, 4z)
B (1x, 3y)
D (1z)
G (4x)
K (4w, 4z)
H (2w, 4y, 3z)
C (2w, 2x, 3y, 3z)
w
Load
ceiling
5
x
Load
ceiling
4
y
Load
ceiling
6
z
Load
ceiling
6

Fig. 15.3

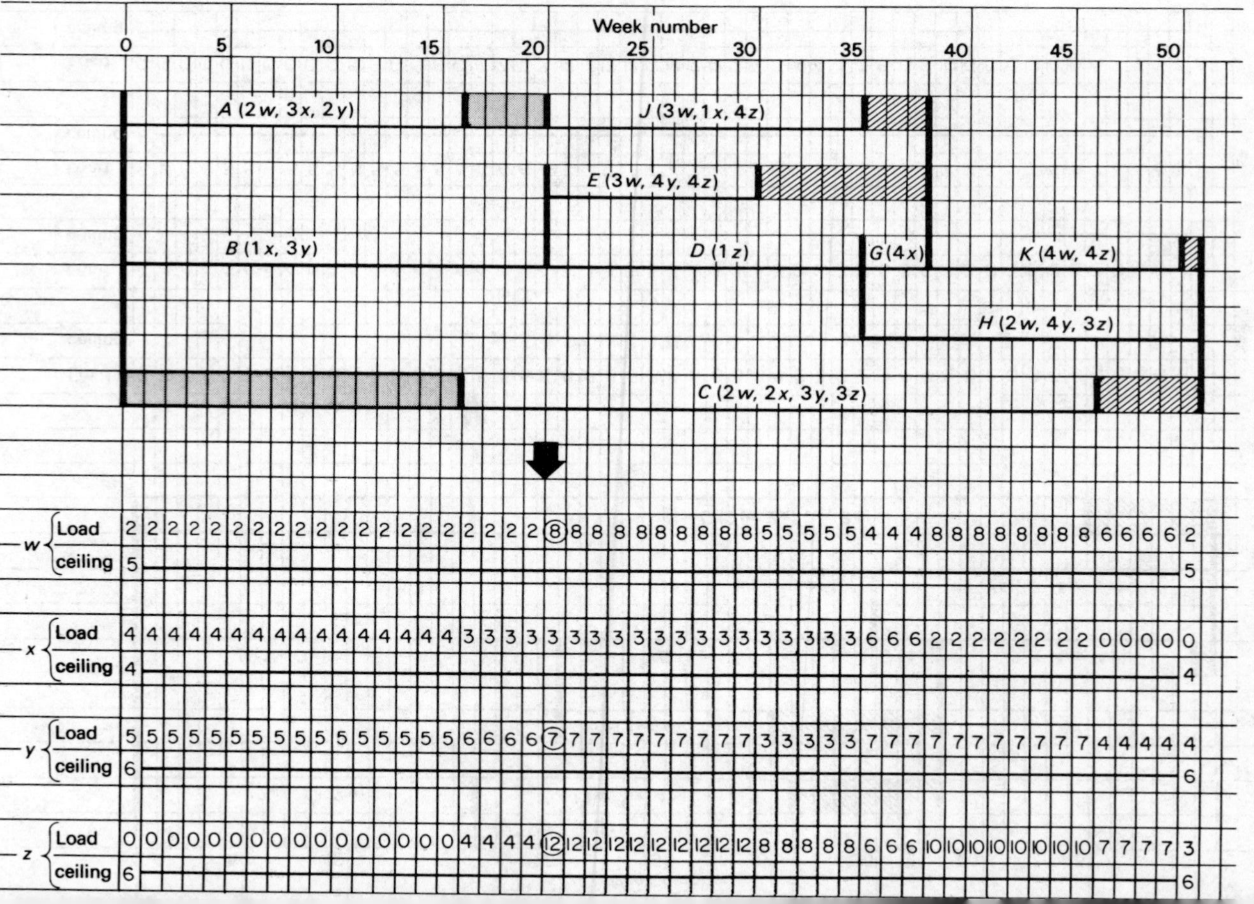
Week number
0 5 10 15 20 25 30 35 40 45 50
A (2w, 3x, 2y)
J (3w, 1x, 4z)
E (3w, 4y, 4z)
B (1x, 3y)
D (1z)
G (4x)
K (4w, 4z)
H (2w, 4y, 3z)
C (2w, 2x, 3y, 3z)
w Load 2 2 2 2 2 2 2 2 2 2 2 2 2 2 2 2 2 2 2 2 (8) 8 8 8 8 8 8 8 8 8 5 5 5 5 5 4 4 4 8 8 8 8 8 8 8 8 6 6 6 6 2
ceiling 5 5
x Load 4 4 4 4 4 4 4 4 4 4 4 4 4 4 4 4 3 3 3 3 3 3 3 3 3 3 3 3 3 3 3 3 3 3 3 6 6 6 2 2 2 2 2 2 2 2 0 0 0 0 0
ceiling 4 4
y Load 5 5 5 5 5 5 5 5 5 5 5 5 5 5 5 5 6 6 6 6 (7) 7 7 7 7 7 7 7 7 7 3 3 3 3 3 7 7 7 7 7 7 7 7 7 7 7 4 4 4 4 4
ceiling 6 6
z Load 0 0 0 0 0 0 0 0 0 0 0 0 0 0 0 0 4 4 4 4 (12) 12 12 12 12 12 12 12 12 12 8 8 8 8 8 6 6 6 10 10 10 10 10 10 10 10 7 7 7 7 3
ceiling 6 6

Fig. 15.4

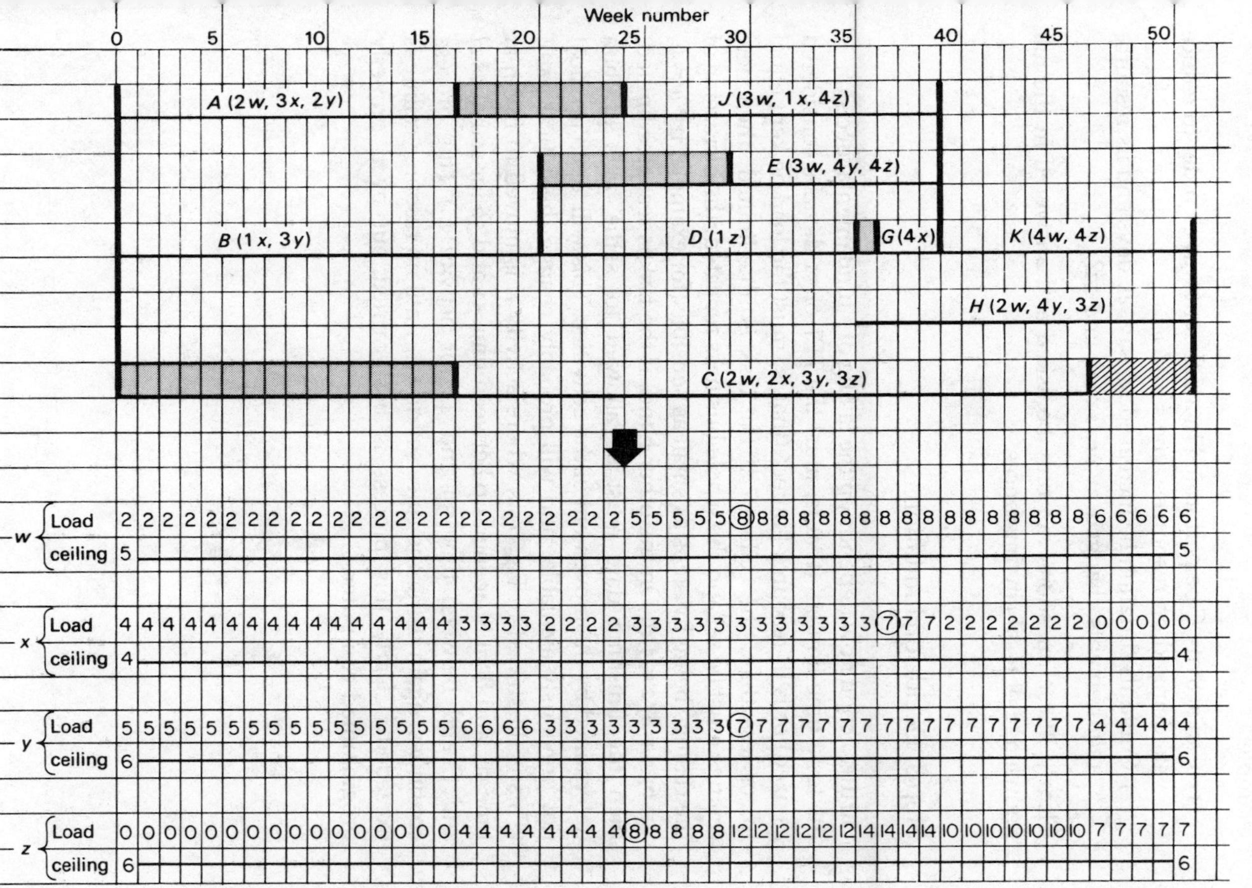

Week number
0
5
10
15
20
25
30
35
40
45
50
A (2w, 3x, 2y)
J (3w, 1x, 4z)
E (3w, 4y, 4z)
B (1x, 3y)
D (1z)
G (4x)
K (4w, 4z)
H (2w, 4y, 3z)
C (2w, 2x, 3y, 3z)
w Load
ceiling 5
x Load
ceiling 4
y Load
ceiling 6
z Load
ceiling 6

Fig. 15.5

impossible the initial policies of the plan should be re-examined. For example:

(1) Can activity *C* be performed in a different way in order to reduce its demand on resources type x and y?
(2) Could its duration be increased to, say, 40 days and its resource requirements changed to $2w$, $1x$, $1y$ and $1z$?

Clearly, these questions can only be asked and answered in the full knowledge of local circumstances.

Step 8. Decide on limitations

Further action now depends on the taking of a fundamental decision: is the task 'time-limited' (that is, must the *TPT* of 51 days be considered as fixed) or is it 'resource-limited' (that is, must the load be kept equal to, or less than, the ceiling)? If the task is time-limited, then new ceilings must be set and the whole business repeated, the increasing of the ceilings being by steps as small as sensible (for example, try $w = 6$, $x = 6$, $y = 8$, $z = 12$). On the other hand, if the task is resource-limited, then the finishing date must be allowed to 'settle' in a position determined by the movement of the activities. This will 'open up' float not previously available, and will probably change the critical path. (This is illustrated in Fig. 15.5 where activity *K* is moved to finish at time 56, and float is opened up between time 51 and 56 for activities *H* and *C* and between time 39 and 44 for activity *E*.) This process continues, again using the rules set down, until no resource ceiling is exceeded. Clearly this process is best carried out by a properly programmed computer.

Resource allocation procedure

1. Draw initial network and determine resources required.
2. Determine resource ceiling.
3. Prepare bar chart in 'all earliest start' position. Inscribe resources required on network.
4. Determine total resource requirements.
5. Establish decision rules.
6. Utilize float to apportion resources.
7. If result unacceptable, re-examine basic logic.
8. Re-examine time/resource limitations.

16 Resource allocation: III An optimum-seeking approach

The resource allocation problem can be resolved by using the 'branch-and-bound' algorithm. It will be demonstrated here by using a very simple problem: in practice, any but the most trivial problem demands a computer with substantial storage capacity. While the method is illustrated here by reference to only one resource, it can be used with as many resources as necessary since it is essentially a 'one step at a time procedure'. Multiple resources make the use of a computer mandatory.

To economize in space only one project network technique (PNT) convention, the activity-on-arrow (AoA) convention, will be used here. The method is identical if the activity-on-node (AoN) display is used. Two basic rules will be followed:

(1) The basic logic of the network is always maintained.
(2) An activity once started will always be finished.

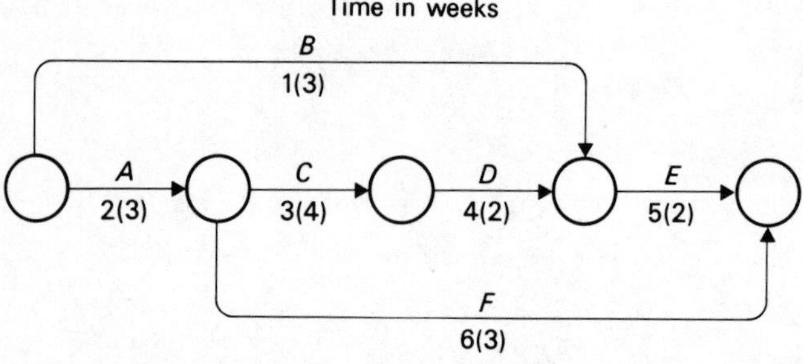

Fig. 16.1 Initial network (resource requirements shown in brackets)

If the initial network (Fig. 16.1) is translated into a bar chart with all activities in the 'earliest start' position, the resource histogram of (Fig. 16.2) results. Here, the resource ceiling of 5 men is exceeded for most of the first 5 weeks, an unacceptable situation. Clearly, not all activities can start as early as possible.

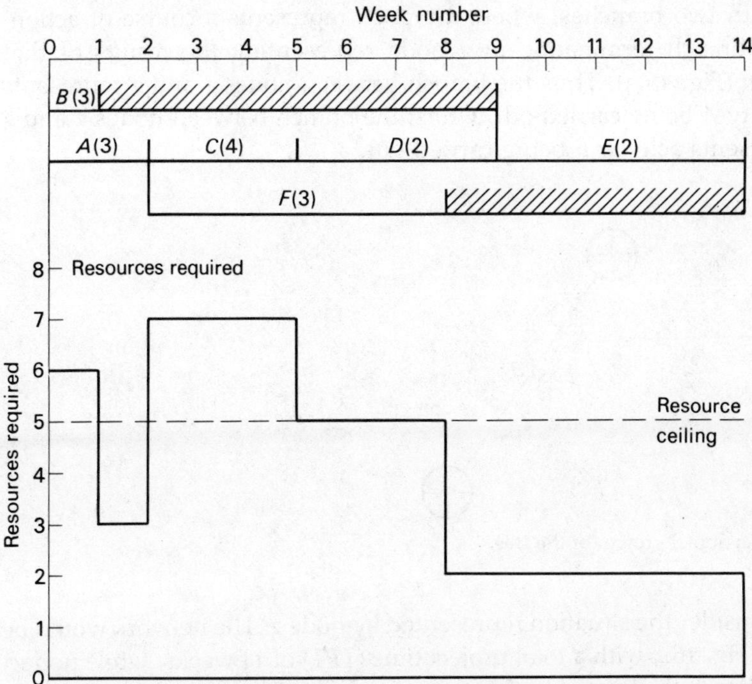

Fig. 16.2 Gantt chart and derived histogram of initial network with all activities starting as early as possible (resource requirements shown in brackets)

Branching

At the outset, three alternatives present themselves:

	Start activity	*Resource requirement (men)*	*Feasible*
1	*A* only	3	Yes
2	*B* only	3	Yes
3	*A* + *B*	3 + 3 = 6	No

The resource ceiling dictates that only the first two alternatives are acceptable ('feasible'), and a 'tree' can 'grow' from an initial node (node

1) with two branches, where a branch represents a course of action, each branch terminating in a node representing the results of that action (Fig. 16.3). Thus the branch between nodes 1 and 2 represents activity *A* being carried out, whilst the branch between nodes 1 and 3 represents activity *B* being carried out.

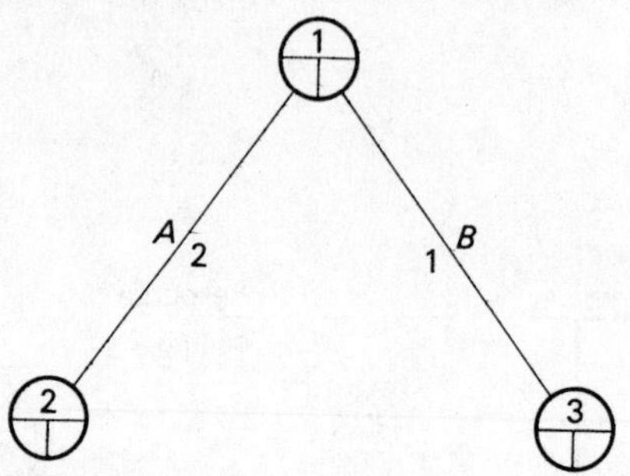

Fig. 16.3 Initial growth of the tree

Consider the situation represented by node 2. The network would be as in Fig. 16.4 with a total project time (*TPT*) of 14 weeks, while node 3 represents the network of Fig. 16.5 with a *TPT* of 15 weeks.

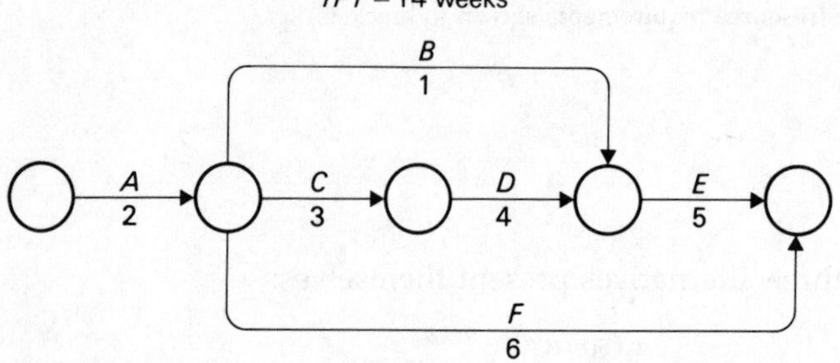

Fig. 16.4 Situation represented by node 2 (*TPT* = 14 weeks)

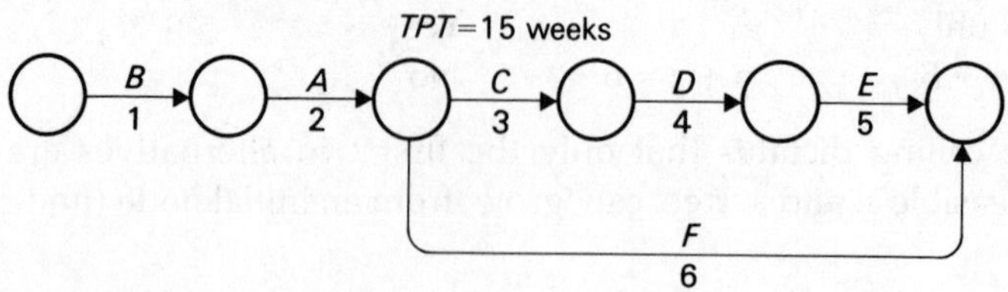

Fig. 16.5 Situation represented by node 3 (*TPT* = 15 weeks)

Bounding

At node 2 the only decision that has been taken is that activity *A* is fixed ('committed') into position with the result that the minimum possible duration time for the project is 14 weeks. This figure can never be reduced no matter how subsequent activities are deployed: indeed it is possible that later resource allocation decisions may cause this time to increase. Since this is the lowest possible *TPT* for this decision it is known as the *lower bound* for this node. Similar considerations show that node 3 has a lower bound of 15 weeks.

It must be emphasized that there is no guarantee that when all resource allocation from a node is complete the *TPT* will equal the lower bound of the originating node, only that the *TPT* from that node can *never be less* than the lower bound.

Conventionally, the 'committed' time at a node (2 weeks for node 2, 1 week for node 3) is entered into the bottom left-hand side of the node, and the lower bound for the *TPT* is entered in the bottom right-hand side, the node number appearing in the top of the node. The committed activity(s) is (are) entered above the appropriate arc its duration time being shown as near to the activity description as possible and below it.

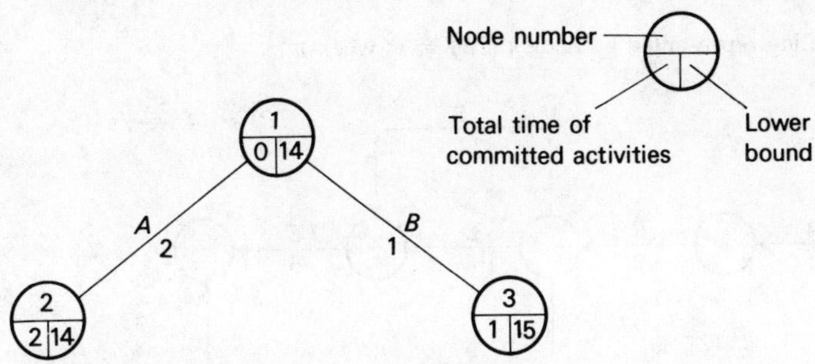

Fig. 16.6 First branches of the resource tree

Further branching and bounding

The resource tree of Fig. 16.6 shows two nodes from which further growth may take place. At this stage the node with the least lower bound is taken as the 'growth bud' since it holds out hope of a lower *TPT*. This promise may need to be tested later.

Node 2 has the least lower bound and 7 actions could follow:

	Commit activity(s)	Resources required (men)	Feasible
1	*B* only	3	Yes
2	*C* only	4	Yes
3	*F* only	3	Yes
4	$B+C$	$3+4=7$	No
5	$B+F$	$3+3=6$	No
6	$C+F$	$4+3=7$	No
7	$B+C+F$	$3+4+3=10$	No

Of these, only the first three are feasible as far as resources are concerned, so that three further branches will grow from node 2, ending in nodes 4, 5 and 6 respectively. Consider node 4. Here activity *A* is already committed (something that is true also of nodes 5 and 6)

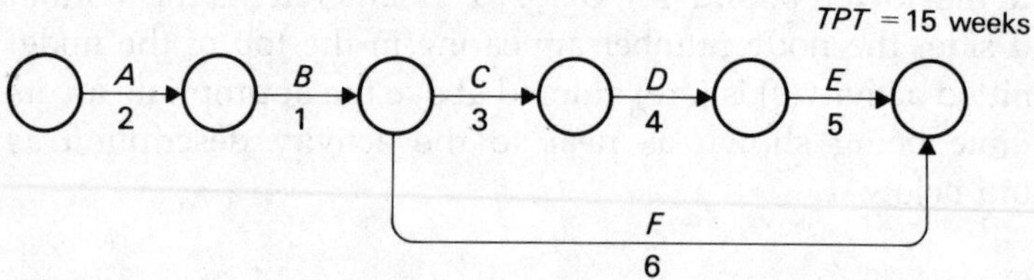

Fig. 16.7 Situation represented by node 4 (*TPT* = 15 weeks)

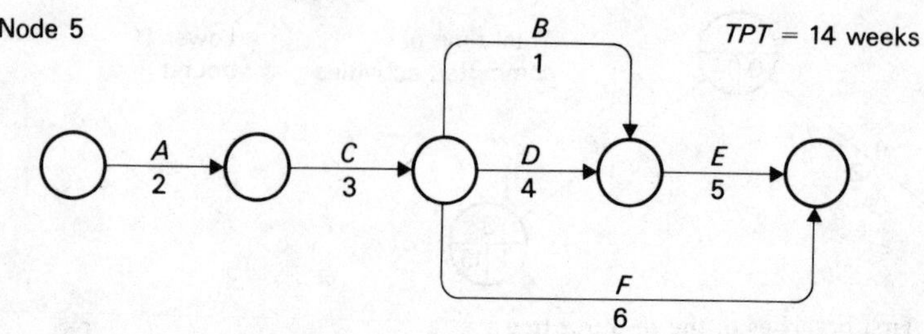

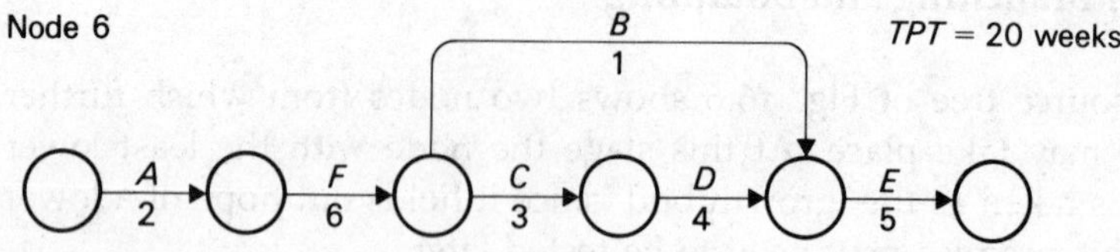

Fig. 16.8 Situations represented by nodes 5 and 6

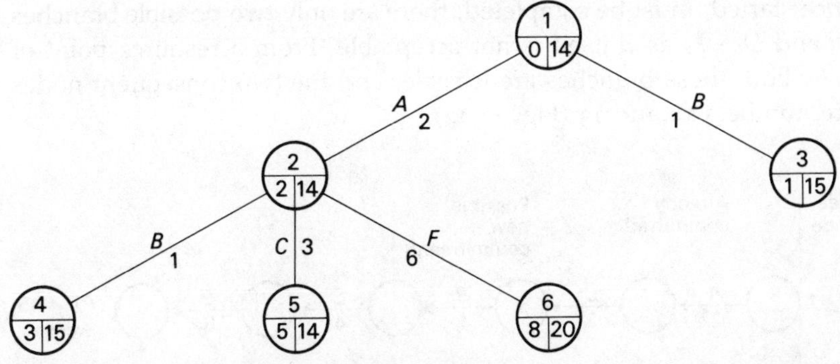

Fig. 16.9 Resource tree at second decision level

and it is followed immediately by activity *B* as in Fig. 16.7. This node, therefore, has a lower bound of 15 weeks and a committed time of 3 weeks. Nodes 5 and 6 similarly are represented by Fig. 16.8. The resource tree has now grown to Fig. 16.9.

Branching again takes place from the node with the least lower bound—node 5—when 7 possibilities present themselves:

	Commit activity(s)	*Resources required (men)*		*Feasible*
1	*B only*		3	Yes
2	*D* only		2	Yes
3	*F* only		3	Yes
4	$B + D_s$	3 + 2 =	5	Yes
5	$B + F_s$	3 + 3 =	6	No
6	$D + F_s$	2 + 3 =	5	Yes
7	$B + D_s + F_s$	3 + 2 + 3 =	8	No

The subscripts *s* and *f* imply the start or finish of an activity. Thus D_s is the start, 1 week, of activity *B* which matches its parallel activity *B*. By the convention initially adopted, if D_s is committed it *must* be followed by D_f in the next commitment of activities, either by itself or in combination with other activities.

Five feasible branches grow from node 5 to nodes 7, 8, 9, 10 and 11. Networks for these 5 nodes are as in Fig. 16.10 and the resource tree grows to Fig. 16.11.

It will be seen that the least lower bound is that of node 10, from which further branching takes place. Remembering that an activity, once started, must be completed, there are only two possible branches D_f and $D_f + F$, as F itself is not acceptable. From a resource point of view both these branches are feasible, and the two consequent nodes are numbers 12 and 13 (Fig. 16.12).

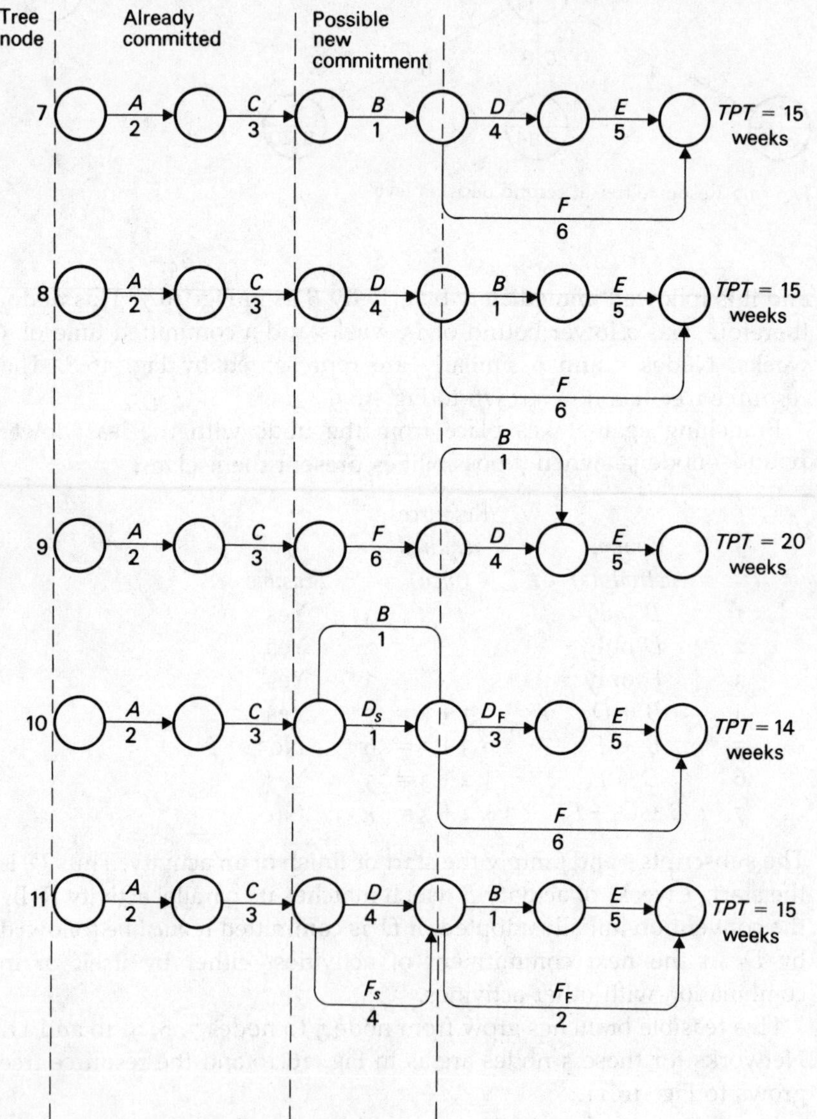

Fig. 16.10 Situations represented by nodes 7, 8, 9, 10 and 11

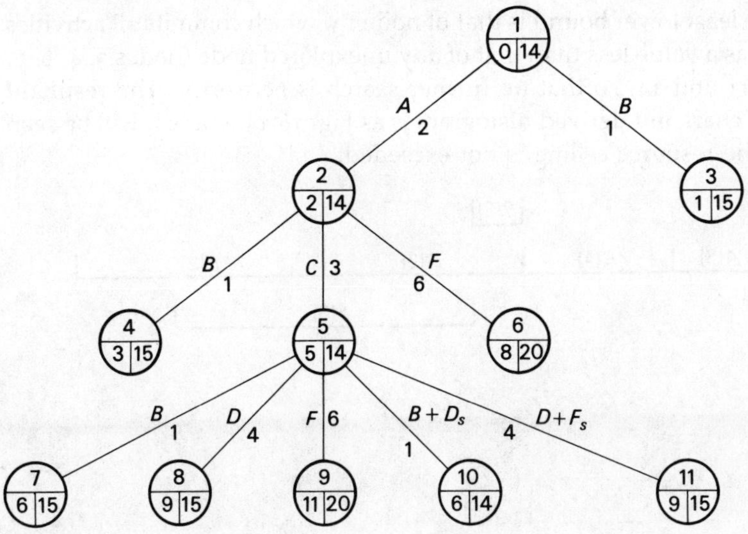

Fig. 16.11 Resource tree at third decision level

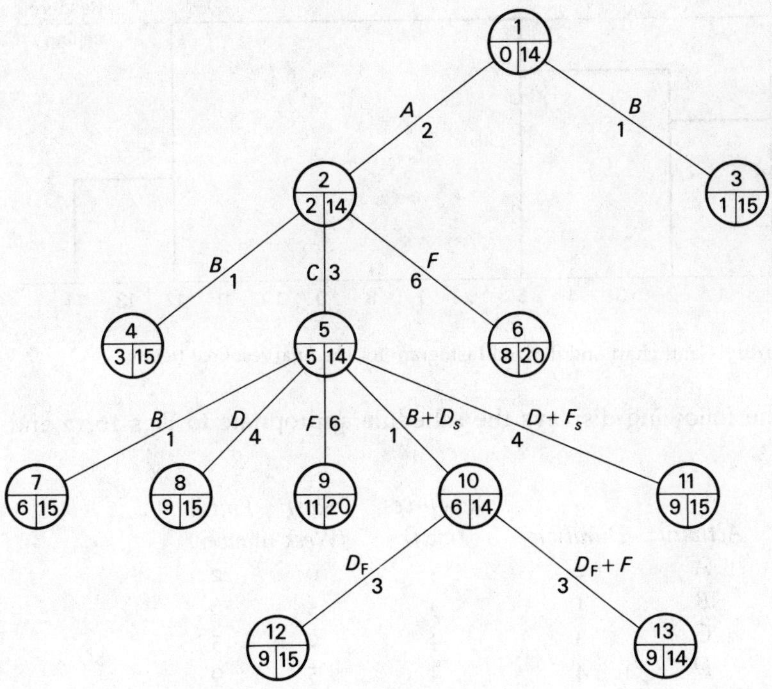

Fig. 16.12 Resource tree at fourth decision level

The least lower bound is that of node 13, which commits all activities and has a value less than that of any unexplored node (nodes 3, 4, 6, 7, 8, 9, 11 and 12) so that no further search is necessary. The resultant Gantt chart and derived histogram is as Fig. 16.13 where it will be seen that the resource ceiling is not exceeded.

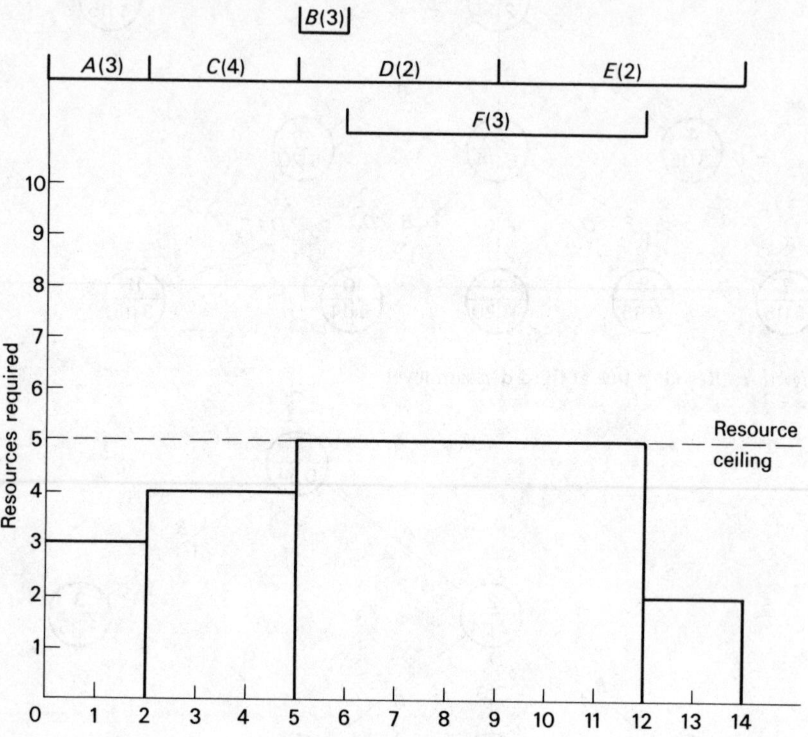

Fig. 16.13 Gantt chart and derived histogram for the final resource tree

The following displays the schedule appropriate to Figs 16.12 and 16.13:

Activity	*Duration*	*Resources (men)*	*Start* (*Week number*)	*Finish*
A	2	3	0	2
B	1	3	5	6
C	3	4	2	5
D	4	2	5	9
E	5	2	9	14
F	6	3	6	12

A table such as the one below may be found useful to summarize the search procedure:

Node number	*Activities committed Description*	*Time*	*Possible start activities*	*Resource requirements*	*Feasible Yes*	*No*	*Generate new nodes*	*Lower bounds*
1	—	—	A	3	√		2	14*
			B	3	√		3	15
			$A+B$	3+3= 6		√	—	
2	A	2	B	3	√		4	15
			C	4	√		5	14*
			F	3	√		6	20
			$B+C$	3+4= 7		√		
			$B+F$	3+3= 6		√		
			$C+F$	4+3= 7		√		
			$B+C+F$	3+4+3=10		√		
5	A, C	5	B	3	√		7	15
			D	2	√		8	15
			F	3	√		9	20
			$B+D_s$	3+2= 5	√		10	14*
			$B+F_s$	3+3= 6		√		
			$D+F_s$	2+3= 5	√		11	15
			$B+D_s+E_s$	3+2+3= 8		√		
10	$A, C, B+D_s$	6	D_f	2	√			
			D_f+F_s	2+3= 5	√			

* Represents the growth bud.

Further searching

The solution of the above example was produced in the *first pass* through the tree. A situation could readily occur when the lower bound at a node was greater than the lower bound of a node at an earlier decision level.

Thus, in the tree of Fig. 16.14 the result of a first pass, at node x, has a lower bound of 35, while the lower bound at an earlier node, node 4, is 30.

The significance of this situation is that node x *guarantees* a feasible solution: further branching from node 4 *may* produce a shorter *TPT*. Whether it is worth the effort and cost of additional branching depends principally upon the difference between the two lower bounds. If the difference is small, the additional gain derived from further bounding must also be small. Equally if the difference is of the order of inaccuracy of the basic data the benefits that are likely to accrue are doubtful.

Two basic search procedures can be identified: one where growth takes place continuously from one node to the next ('the latest active

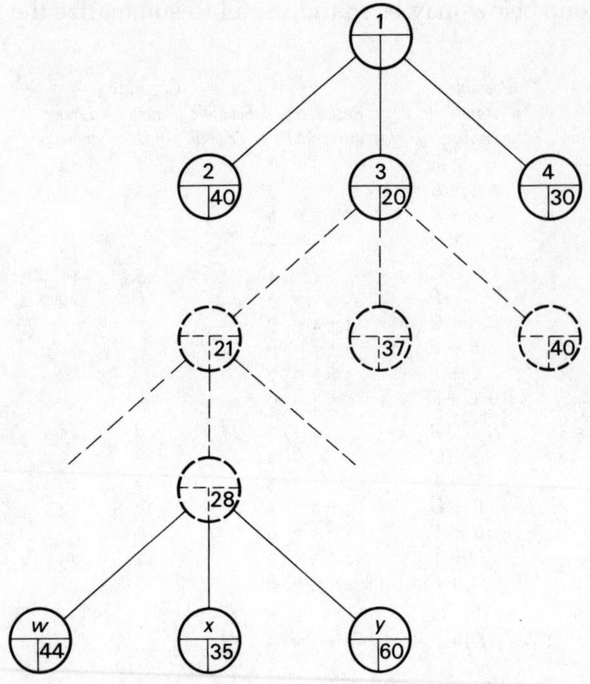

Fig. 16.14 Resource tree where the first pass solution exceeds an earlier lower bound

node procedure'), reference only being made to earlier node when the first pass is completed. The other method requires a continuous search across all nodes at all times ('frontier search') where branching takes place not necessarily from the most recent node but from whichever node, wherever located, with the least lower bound.

In general, it will be found that the latest active node procedure is the most economical on storage space, and in PNT the first pass solution is likely to be acceptable.

17 Control and project network techniques

Control is here used in a special sense; it is *not* used to mean supervision or direction, but it *is* used to mean the comparing of what actually takes place with what has been planned to take place. As such, it has the significance given to control in budgetary control, where the actual expenditure is compared with the planned expenditure, and any differences (the accountant's 'variances') are reported to the person responsible for the expenditure.

The essential features of control

An industrial control system appears to have six essential features:

(1) A plan must be made.
(2) This plan must be published.
(3) Once working, the activity being controlled must be measured.
(4) The measurements must then be compared with the plan.
(5) Any deviations must be reported to the appropriate person.
(6) A forecast of the results of any deviations must then be made, and corrective actions taken to cause the activity to continue in a way that will produce the originally desired result, or, if this is not possible, a new plan must be made.

The above six features appear to be general to any organizational control system, and they should therefore be considered when using project network techniques (PNT) as a control technique.

Planning and publishing

Looking at PNT it will be seen that inherently it contains the first two basic features, planning and publishing, most adequately. In many tasks it is the only possible planning technique available, and its use as a means of communication has already been commented upon; it is an excellent means of publishing a plan.

Measuring

Measuring activity is not an integral part of PNT, although the discipline imposed by constructing the network, and the consequent depth of insight into the project, will indicate how best these measurements can be made. There are, again, a number of general features of control measurements that have emerged from other control situations which appear to apply when using PNT in a control situation.

General features of control measurements

(1) The measurement should be appropriately precise. Any measurement can be increased in precision by an increase in the cost of making the measurement. PNT indicates very clearly which activities need to be precisely measured (those on the critical path) and those which do not need such a high precision. For example, in the network already discussed, measuring activity *B* to the nearest day could well be useful, since it is on the critical path and any 'slip' would result in an increase in overall time for the project. On the other hand, activity *C* has a total float of 21 weeks, and to monitor this to the nearest day would be uselessly expensive.

(2) The measurements should be pertinent. This is quite self-evident, yet the files of industry bulge with data that have been collected and that are not used. It is essential to question the use that will or can be made of any data.

(3) The speed of collection of the information must be rapid compared with the time-cycle of the system as a whole. In a project lasting two years, collecting information and processing it every two weeks is probably adequate, since it will allow corrective action to be taken. No general rule can be laid down here, but it must be remembered that, as a project progresses, the time remaining for completion diminishes and, hence, the speed of collection may need to increase. Thus it is quite usual, at the outset of a long project, to receive reports every month but, as time advances, to reduce the reporting time to once a fortnight, later to once a week and eventually to once a day. The essential thing to remember is that measurements must be taken frequently enough to allow useful action to be taken. Collecting progress information is both difficult and expensive.

(4) Measurements need to be accurate or of consistent inaccuracy. As with the degree of precision, so with accuracy; accuracy can be

bought with increased cost. It is frequently cheaper to accept a measuring technique that is known to be inaccurate but consistent than to attempt to obtain a very high accuracy. Consistent inaccuracies can be allowed for; high accuracy inevitably results in increased cost. Here again, PNT indicates where a high accuracy measurement should be made and where a low accuracy one is tolerable. Thus, activity *B* has little float, and 10 per cent accuracy, whereas activity *C* which has considerable float, could probably tolerate a very much greater error—say 50 per cent in the early days of the project. (*Note:* Although accuracy and precision are related, they are quite separate concepts, and they should not be confused.)

(5) The number of data processing points should be kept as small as possible. Once a measurement is made, it should be passed through as few processing departments as possible. Not only will handling delay the using of the information, but it will inevitably cause distortions that are very difficult to eliminate.

Comparing and reporting

Some of the ways in which comparing and reporting may be carried out with PNT are noted below. While it is possible to devise many other methods, it is sensible to avoid letting the ingenuity of the method become an end in itself. The author has observed many clever techniques in which the mechanics of the method have obscured the results. *The simplest method is always the best.*

1. The network itself

As work progresses, the network itself can be marked in some way to indicate that work has been done, although this tends to be unsatisfactory since the location of an activity is not related to the time when it should be performed. Another method in arrow-on-arrow (AoA) is to use the quadrated node symbol, and insert the 'arrival time' in the top segment. Thus, if activities 2–8, 3–8 and 7–8 are all completed by time 39, 39 is inserted in the circle (Fig. 17.1). Providing the inserted time is equal to or less than the right-hand time, then the project as a whole can be achieved without re-planning.

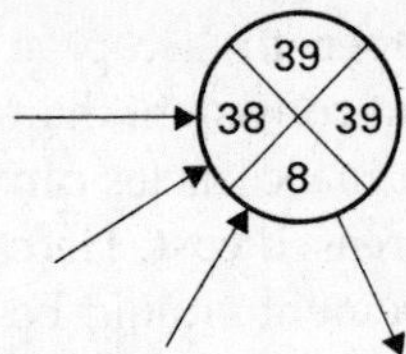

Fig. 17.1

2. The bar chart

If the arrow diagram has been translated into a bar chart, performance can be signalled by drawing a progress bar (see p. 2). This is both simple and graphic, and in many cases is the most appropriate way of displaying progress information.

3. Re-analysis

In complex projects it is virtually impossible to represent the whole situation graphically. Under these circumstances, PNT can prove to be of inestimable value. By taking the original network and inserting into it the *actual* times, instead of the expected duration times, it is simple to re-analyze the network and see the effects of the actual work. These effects will be shown up most clearly by a change in float. For example, assume that at the end of week 30 the situation with regard to the sample network is as follows:

	Duration time		
Activity	*Expected*	*Actual*	*Notes*
A	16	18	Complete
B	20	19	Complete
C	30		Not started
J	15	20	Partly complete; re-estimate
D	15	15	Partly complete; re-estimate
E	10	15	Partly complete; re-estimate
G	3		Not started
H	16		Not started
K	12		Not started

Note that activity *C* has a latest start time (*LST*) of week 21, and this has already been exceeded. It is therefore necessary to take cognizance of this fact, which is done here by inserting another activity, *L*, which is a delay activity of duration 30 weeks. Another way of showing this would be by increasing the duration time of activity *C* to 60 weeks.

Analyzing the network as previously explained will give the following result:

Activity	Duration	Start time Early	Start time Late	Finish time Early	Finish time Late	Float Tot.	Float Free	Float Ind.
A	18	0	10	18	28	10	0	0
B	19	0	10	19	29	10	0	0
L	30	0	0	30	30	0	0	0
C	30	30	30	60	60	0	0	0
J	20	18	28	38	48	10	0	0
D	15	19	29	34	44	10	0	0
E	15	19	33	34	48	14	4	0
G	3	34	45	37	48	11	1	0
H	16	34	44	50	60	10	10	0
K	12	38	48	50	60	10	10	0

This shows quite clearly the effect of the various performances, the principal one being that the critical path has now shifted to start–*L*–*C*–finish, the float having been opened up in the previous critical path. The project can now be re-examined to see where and how this slipping can be made good.

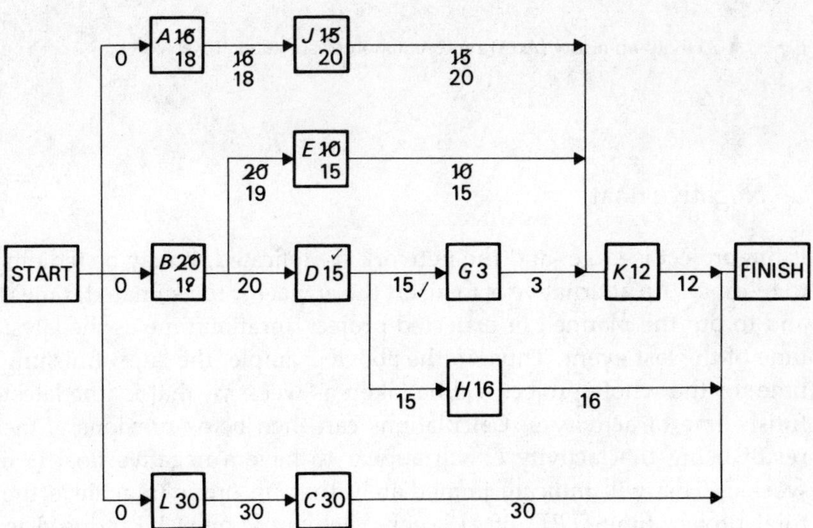

Fig. 17.2 Activity-on-node (AoN) representation of situation at week 30

An alternative technique is to re-draw the network, leaving out those activities that are complete, and then re-analyzing in the normal way, substituting the actual date for the date of the first event. This will have the effect of gradually collapsing the network from the left, and in large networks a progressive simplification will result. It will obviate the need for inserting new 'delay' activities. In the example shown there is little advantage in doing this.

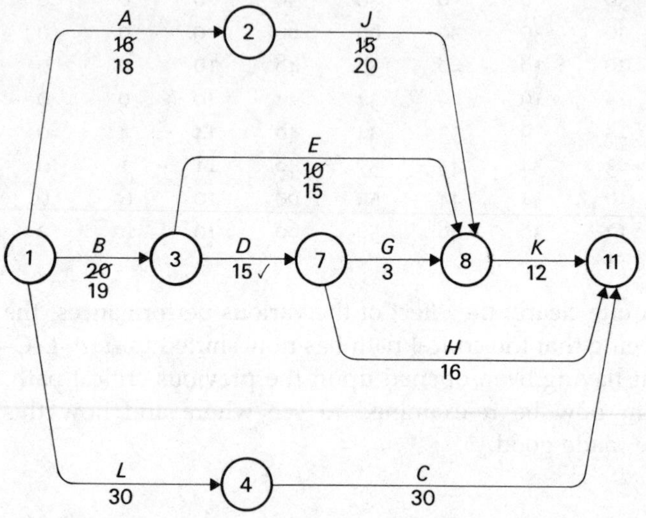

Fig. 17.3 Activity-on-arrow (AoA) representation of situation at week 30

4. Negative float

If the project is large, and the network complicated, it may be tedious to re-draw. An alternative is to insert the actual (or re-estimated) times, and to put the planned or expected project duration time as the latest time of the last event. Thus, in the above example, the latest finishing time for the whole project will be taken as week 51, that is, the latest finish time of activity *L*. Calculations can then be as previously, the result being that activity *C* will appear to have a negative float (−9 weeks). This will indicate immediately that, in order to achieve the total project time (*TPT*) of 51 weeks, activity *C* must be reduced in duration time by 9 weeks. In many cases this negative-float technique is most appropriate, since it can be obtained quickly by re-processing the data through a computer.

5. Time remaining and the elapsed time arrow

To reduce the work involved in analysis, insert into the network the *time remaining* for the various activities, discarding those activities that are completed (that is, those where the time remaining = 0). Thus, using the example of section 3 above, the times remaining would be:

Activity	*Time remaining*
A	0
B	0
C	30
J	8
D	4
E	4
G	3
H	16
K	12

and the network would be as shown in Fig. 17.4 or 17.5,

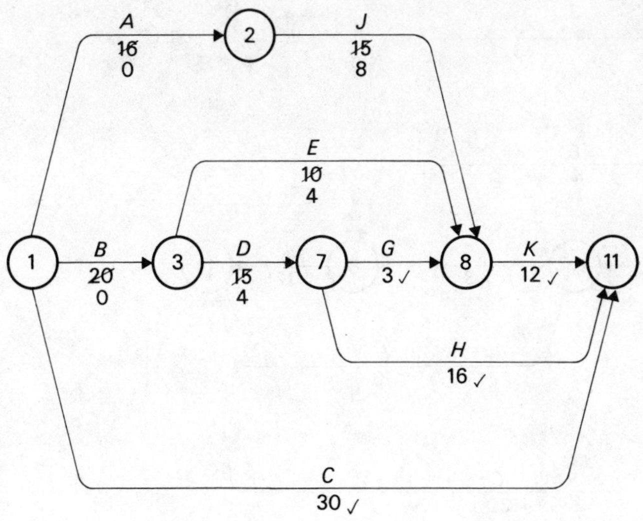

Fig. 17.4 AoA progress network using 'time remaining'

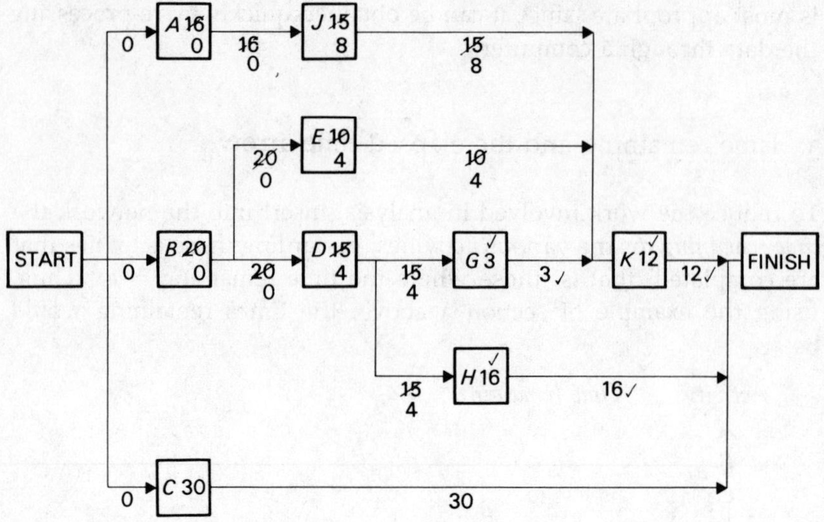

Fig. 17.5 AoN progress network using 'time remaining'

or, discarding the zero time activities, the network would shrink to that given in Figs 17.6 and 17.7.

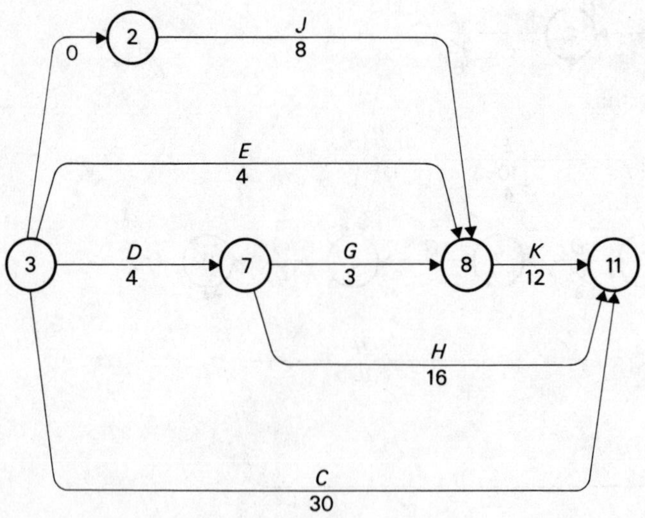

Fig. 17.6 AoA 'time remaining' network

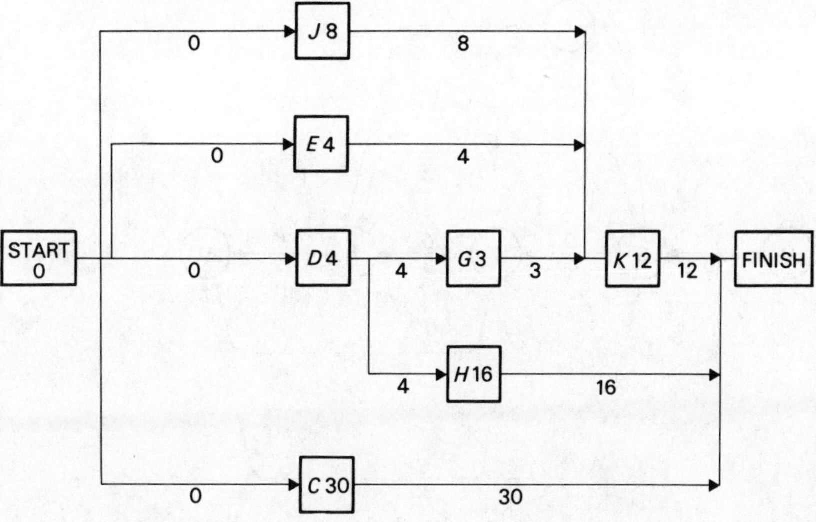

Fig. 17.7 AoN 'time remaining' network

This is a smaller network to handle than the original, but its float times would be identical to those shown in the table in section 3. However, the *TTP* would be shown to be 30 weeks, and it would not be possible to compare directly these activity times with the activity times of the original analysis which were based on a *TPT* of 51 weeks. This disadvantage can be overcome by inserting an *'elapsed time'* arrow at the beginning of the network, the duration time of this being equal to the time that has elapsed since the start of the project. In the example so far used the elapsed time arrow would have a duration of 30 weeks (since the project is at the end of week 30) and the network would become as illustrated in Figs 17.8 and 17.9.

Figures 17.8 and 17.9, being analyzed, would give results directly comparable with the original analysis (that on p. 78) and any comparisons required would easily be made. Ideally, of course, the duration of the elapsed time arrow at the conclusion of the project should be equal to the original project-time.

The computer is of particular value in 'updating' since it is happy to carry out repeated calculations. *Note:* In any form of re-analysis it is essential, in AoN, to modify not only the duration times but also *all* dependency times.

One continually recurring problem in 'updating' networks is the difficulty in obtaining useful statements from the operating point. It is

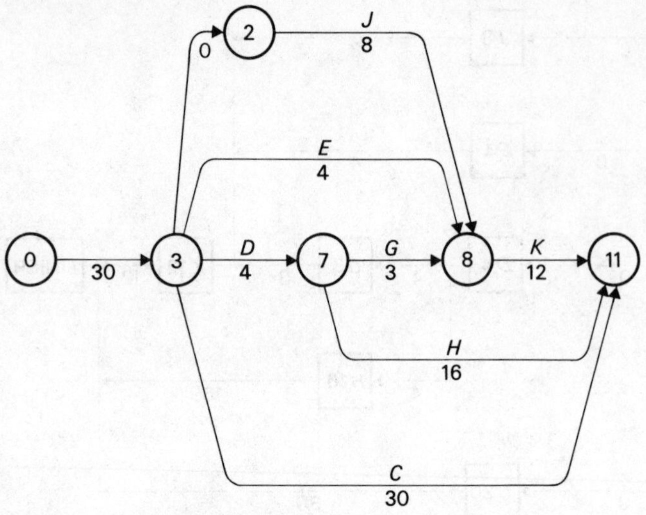

Fig. 17.8 AoA 'time remaining' network with elapsed time activity

impossible to give a simple solution to what is a very complex problem. Two things can be said:

(1) Avoid recrimination. If an activity over-runs, be careful not to use this as the opportunity to create a fuss. The past is dead—take steps to *avoid* a recurrence of whatever failed. A respected

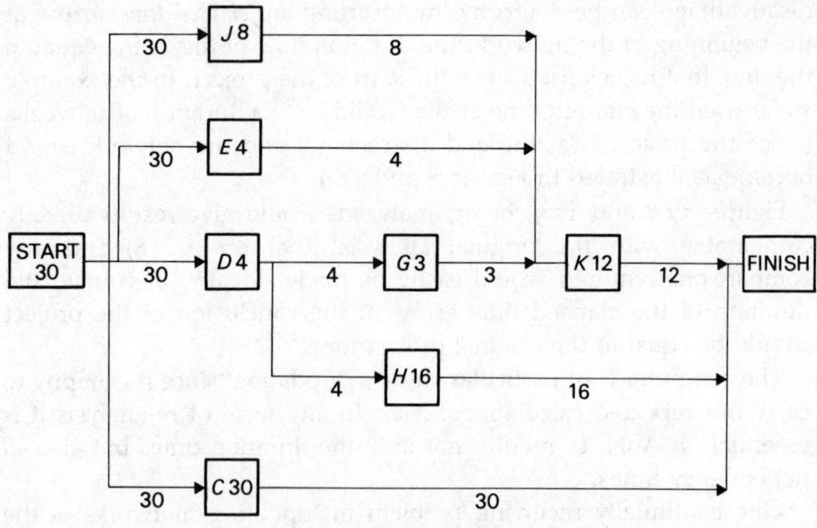

Fig. 17.9 AoN 'time remaining' network and elapsed time activity

networker once referred to PNT as a 'do-it-yourself hangman's kit'. Let PNT be a tool to assist, not a weapon to assault.

(2) Progress should be reported in the form 'not complete' or 'complete'. Thus, to the question:

> 'Is activity *X* complete?'
> A response should either be 'Yes' or 'No' and if 'No' then a second question:
> 'How much time is required to complete activity *X*?'
> should be asked. Statements such as: 'nearly finished', 'almost finished', 'just a little to do', 'it'll soon be done!' and so on should be eschewed. They both inculcate a sloppiness of mind and an avoidance of responsibility.

The author has found that a useful way of progressing work is by employing the *LST*s and *LFT*s derived from the analysis of the network. Thus, it is possible by scanning the analysis to determine for a number of weeks in the immediate future which activities 'must' be started and which activities 'must' be finished. At progress meetings it is then possible to identify exactly which activities must have been started by the date of the progress meeting and which must have been finished. This sorting of activities into 'must start' 'must finish' categories is very easily carried out by means of a computer.

Forecasting and taking corrective action

When performance does not conform with plan, and it is necessary to take corrective action, it must be clearly understood that PNT does not remove any responsibility from the manager concerned; indeed, by causing areas of authority to be clearly distinguished, it reinforces and emphasizes the manager's position. PNT is neither a prophylactic nor is it a panacea, and any failures to achieve an agreed plan must not be laid at the door of PNT; they will rest, as always, with the manager.

This having been said, it must be pointed out that PNT has a particular use in this field, namely that it will enable predictions of resultant actions to be deduced from present or past action. For example, any 'slip' in a critical activity will result in the whole project 'slipping'. To correct this it may be possible to transfer labour from other non-critical activities, and the consequent effects of this can be clearly seen by considering the arrow diagram, or re-passing the modified data through a computer if one is being used. This predictive value of PNT is probably unique among planning systems, and certainly is of great value in real-life situations.

PNT and other control systems

All the previous discussions have considered PNT in isolation: this arises from the restrictions imposed by writing a text in general terms. The data used in, and the information derived from, PNT should all be integrated with other control systems. The time taken to perform an activity is of interest not only to the PNT planners but also to the costing department, the wages department, the material-control section, and so on. Failure to co-ordinate the work of various departments will lead to dissipation and duplication of effort, and will also prevent those cross-checkings that can do so much to make up for inadequacies. Not only this—a number of imperfect systems acting in parallel can often produce a total accuracy, which one highly perfected system cannot attain, and this can enable economies to be made that do not result in any degrading of information. It is therefore highly desirable that PNT be used *as a control technique* as well as a planning technique, and that its work here should be integrated as closely as possible with all other management controls. This is discussed further in Chapter 18.

The features of control
and control measurements

1. Plan
2. Publish
3. Measure
4. Compare
5. Report
6. Forecast and correct

(Publish, Measure, Compare:)
(1) Appropriately precise
(2) Pertinent
(3) Fast
(4) Accurate/consistent
(5) Minimum handling

18 Cash as a resource

The availability of cash, like the availability of labour, equipment or space, can be a constraint upon the achievement of a project. There is, however, one factor which, in general, sets it apart from labour—namely that of time.

A statement that 'the labour resource ceiling is 5 men' is usually an abbreviated form of a more correct statement:

> 'The labour resource ceiling for the duration of the project (or for some portion of the project) is 5 men'

and this implies that 5 men are available throughout the specified period—they can be used first on one activity and later on others. Cash, however, is exhausted once it is used—'£1,000 is available' means, in fact, that once £1,000 is used, it is no longer available. Furthermore, it need not be present until it is required—it may be in a bank or some other store, available at call but (hopefully!) not generating cost. The £1,000 may be on deposit in a bank generating interest, or part of an overdraft facility, only generating cost when it is actually used. The difference between these two types of resource is recognized by the names '*non-storable*' and '*storable*' resources, labour being typically non-storable, cash being typically storable.

The storable and exhaustible character of cash means that a histogram is not the appropriate measure and display of cash. Thus, consider two activities, *A* and *B*, which immediately follow each other and where *A* involves an outlay of £600 while *B* involves an outlay of £400. A histogram such as Fig. 18.1 clearly does not represent the situation. By time *t*, assuming that the cash is engaged at the *start* of each activity, £(600 + 400) = £1,000 have been committed, and a more useful diagram would be as given in Fig. 18.2.

Note: The illustration used assumed that the whole of the cash for an activity was committed at the start of the activity. Clearly, if it were committed in some other way—for example, linearly with time—Fig. 18.2 would be different, but the *accumulative* nature of the process would remain.

Figure 18.2 is a *cash flow* diagram showing how the accumulated cash flows out of the organization. In some situations, where part payments

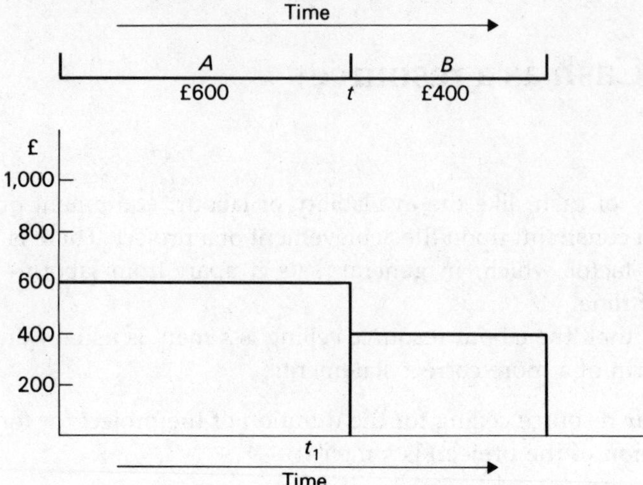

Fig. 18.1

are made, for example, cash can flow *into* the organization and the cash flow diagram can drop by the amount of the inflow.

Note: Financial managers are more likely to draw Fig. 18.2 in an inverted form, showing the cash as a negative value since it is an *outflow*. Project planners tend, on the other hand, to use the form as shown, since it is used later as a convenient control chart.

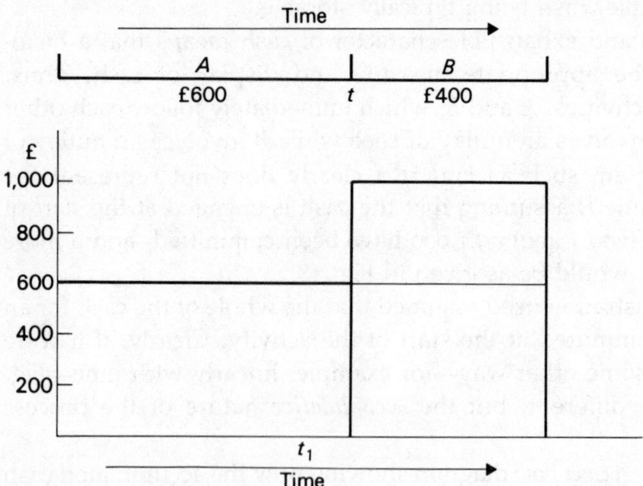

Fig. 18.2

Cash flow diagram

The direct costs involved with the basic network are assumed to be:

Activity	*Duration (weeks)*	*Cost (£)*
A	16	3,200
B	20	1,000
C	30	6,600
D	15	600
E	10	2,200
G	3	900
H	16	800
J	15	1,500
K	12	3,600
	Total	20,400

This sum of £20,400 is the *total* expenditure on the project: it is not a particularly useful figure for control purposes since it is only reached at the conclusion of the project, and deviations revealed at this time are of 'historical' interest only—no correcting action can be taken.

A more useful measure is the outflow of cash with time, allowing comparison of the actual outflow with a planned outflow. Assuming that the flow of cash is constant with time across each activity (the 'spend rate' is constant), then:

Activity	*Duration (weeks)*	*Effective weekly cost (£)*
A	16	200
B	20	50
C	30	220
D	15	40
E	10	220
G	3	300
H	16	50
J	15	100
K	12	300

Assume also that, for reasons of no consequence here at the moment, the project has been scheduled in the following way:

Activity	Duration	Start (week number)	Finish (week number)
A	16	0	16
B	20	0	20
C	30	16	46
D	15	20	35
E	10	20	30
G	3	35	38
H	16	35	51
J	15	16	31
K	12	38	50

(*Note*: As is usual when a project is scheduled, float has effectively disappeared.) The above situations can be displayed on the Gantt chart

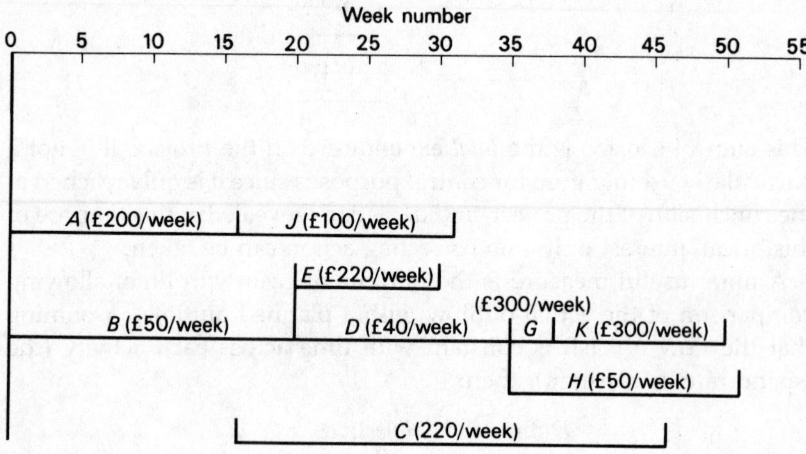

Fig. 18.3

of Fig. 18.3 from which the cash flow can be calculated as below:

Week number	Activities involved	Weekly outgoing (£)		Cumulative outgoing (£)
1	*A B*	200 + 50	= 250	250
2	*A B*		250	500
3	*A B*		250	750
4	*A B*		250	1,000
5	*A B*		250	1,250
6	*A B*		250	1,500

Week number	*Activities involved*				*Weekly outgoing (£)*	*Cumulative outgoing (£)*
7		*A*	*B*		250	1,750
8		*A*	*B*		250	2,000
9		*A*	*B*		250	2,250
10		*A*	*B*		250	2,500
11		*A*	*B*		250	2,750
12		*A*	*B*		250	3,000
13		*A*	*B*		250	3,250
14		*A*	*B*		250	3,500
15		*A*	*B*		250	3,750
16		*A*	*B*		250	4,000
17	*J*		*B*	*C*	100 + 50 + 220 = 370	4,370
18	*J*		*B*	*C*	370	4,740
19	*J*		*B*	*C*	370	5,110
20	*J*		*B*	*C*	370	5,480
21	*J*	*E*	*D*	*C*	100 + 220 + 40 + 220 = 580	6,060
22	*J*	*E*	*D*	*C*	580	6,640
23	*J*	*E*	*D*	*C*	580	7,220
24	*J*	*E*	*D*	*C*	580	7,800
25	*J*	*E*	*D*	*C*	580	8,380
26	*J*	*E*	*D*	*C*	580	8,960
27	*J*	*E*	*D*	*C*	580	9,540
28	*J*	*E*	*D*	*C*	580	10,120
29	*J*	*E*	*D*	*C*	580	10,700
30	*J*	*E*	*D*	*C*	580	11,280
31	*J*		*D*	*C*	100 + 40 + 220 = 360	11,640
32			*D*	*C*	40 + 220 = 260	11,900
33			*D*	*C*	260	12,160
34			*D*	*C*	260	12,420
35			*D*	*C*	260	12,680
36		*G*	*H*	*C*	300 + 50 + 220 = 570	13,250
37		*G*	*H*	*C*	570	13,820
38		*G*	*H*	*C*	570	14,390
39		*K*	*H*	*C*	300 + 50 + 220 = 570	14,960
40		*K*	*H*	*C*	570	15,530

Week number	*Activities involved*	*Weekly outgoing (£)*		*Cumulative outgoing (£)*
41	K H C		570	16,100
42	K H C		570	16,670
43	K H C		570	17,240
44	K H C		570	17,810
45	K H C		570	18,380
46	K H C		570	18,950
47	K H	300 + 50	= 350	19,300
48	K H		350	19,650
49	K H		350	20,000
50	K H		350	20,350
51	H	50	= 50	20,400

This gives the 'cash flow' diagram of Fig. 18.4.

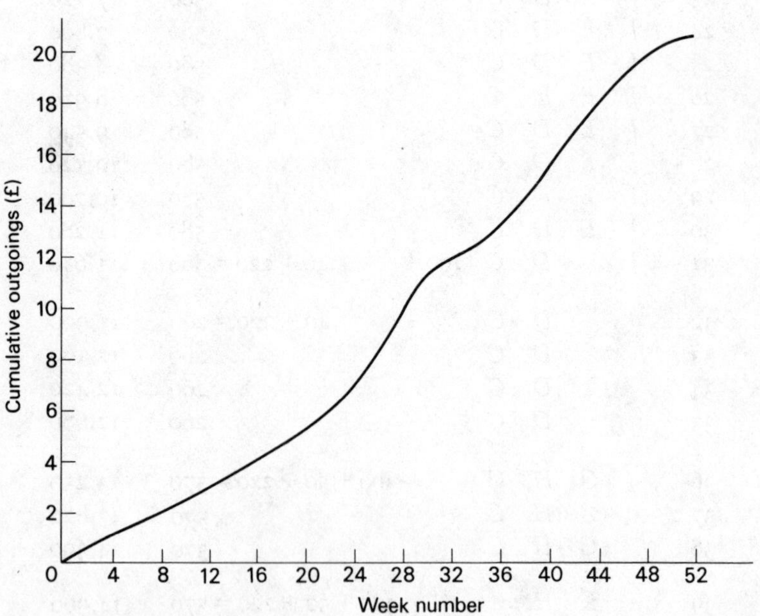

Fig. 18.4 Scheduled cash flow

Cost control

The cash flow calculations above show the *planned* situation. To be able to exert control it is necessary to compare the actual situation with the plan. It is not sufficient, however, to record that at week 20 the actual outgoings were £6,000, the planned outgoings being £5,480. This apparent over-expenditure could have come about by weekly costs being in excess of planned costs—for example, the actual costs of activity *B* could have been £76 a week instead of £50 a week—or more work could have been done than had been anticipated—for example, just over 2 weeks of activity *E* had been completed, or a combination of different costs and work achievements had occurred.

To resolve the above difficulty three figures are required:

(1) Planned expenditure.
(2) Actual expenditure.
(3) Value of work completed.

The cash flow diagram will then appear as in Fig. 18.5, where an overspend at week *X* is accompanied by an under-achievement at that time. The 'value of work completed' figure is obtained from the initial budgeted figures, that is, if, for example, activity *B*, budgeted at 20

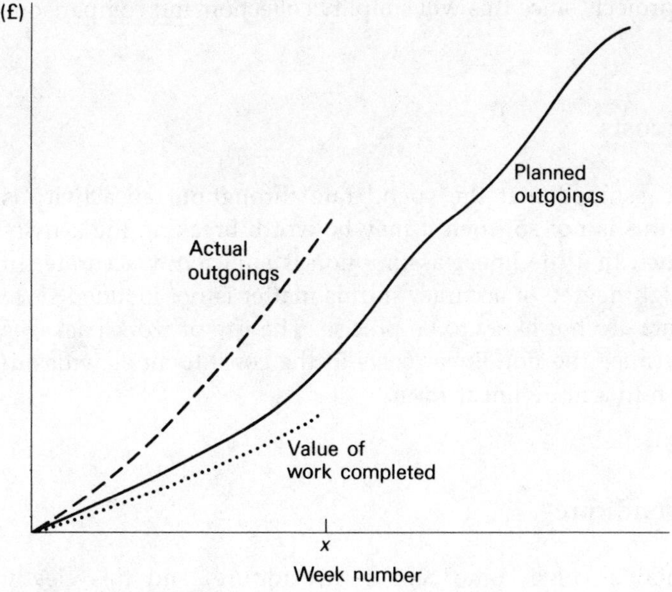

Fig. 18.5 Cost control diagram (situation at week *X*)

weeks at £50 a week is completed in 20 weeks at £100 a week, the value is shown at the original £20 × 50 = £1,000. Equally, if it were *completed* in 15 weeks at £50 a week, the value would still be recorded at £1,000.

Work packages

A network may comprise a much larger number of activities than there are cost centres, or it may be that the collection of costs in the same detail as the work detail of the network is unnecessary. In this type of situation *work packages* may be defined, a work package being a collection of activities brought together for administrative usefulness. Thus, activities *D*, *E* and *G* might constitute a single work package, and within this there might be a number of different trades, each with its own costs. Reporting on this work package might then consist of reporting on the work of the constituent trades, both single and accumulated. Much ingenuity has been expended on devising cost code systems that permit easy reporting and identifying of costs. Cost control usually demands the use of a computer, and the instructions for the computer program being used should be consulted when adopting a cost code. Once this code is adopted it should be used over all possible projects since this will simplify collection and comparison.

Non-linear costs

It has been assumed that the spend rate throughout an activity is constant. If this is not so, then it may be worth breaking the activity into parts such that the linear assumption is sufficiently accurate. In general, a high degree of accuracy in this matter is not justified since the initial data are not likely to be precise. The use of work packages will tend to cause the non-linear costs in the constituent activities to average out into a more linear form.

Capital expenditure

A project may involve some capital expenditure, and this clearly cannot constitute an expenditure over time—a substantial sum is released at one instant. Possibly the most satisfactory device here is to introduce a new activity ('pay for . . .') with a nominal duration and located in position by a real-time dummy. The location of the capital expenditure must be a joint decision of all concerned and not a unilateral one by the project manager.

Indirect expenses—the use of hammocks

All the costs so far discussed have been direct costs, that is, they have been assignable directly to activities or tasks. There are usually other costs that are not directly assignable in this way, the *overhead* costs. The burden of these overheads are supported by *hammock* activities, which are inserted into the network. For example, if activities A and J derive from the same cost centre, and the overhead rate for that centre is £y/week then a new activity, the hammock, is inserted into the network from the tail of the first activity to the head of the second (or last, if there are a number of activities), whose duration is equal to the total activity span of all activities being so 'hammocked'. Cost contributions can then be calculated as above.

Time and cost control

The considerable use made of time control by networks is largely due to the simplicity of the technique. The value of adding cost to the time control is undoubted. Unfortunately, the difficulties encountered in practice are considerable; they are of two kinds:

(1) Substantially more data have to be processed and stored, and their manipulation is invariably tedious. This problem can be readily overcome by the use of a computer, indeed the author has met no situation where cost control has been seriously undertaken manually.
(2) Costs are often extracted and recorded for elements unlike the network activities, if indeed they are extracted and recorded at all. Furthermore, managerial decisions concerning the apportioning of overheads are often not made, or made in a form quite inappropriate for networking.

Despite these setbacks, time-cost control is highly desirable, and on major projects it may well be worth while carrying out the necessary assault on these long-established procedures that are so strongly defended.

The time value of money

The value of money depends upon its availability, £1 available *now* being worth more than £1 available *later*, since the 'present' money can earn some return.

Present value (*PV*)

Assume that £1 is available now, and that it can be immediately invested to produce an annual income of 10 per cent. The £1 would grow as follows:

		Value (£)
Now, beginning of year 1	1	1
End of year 1	$1 + 1 \times 0.10$	1.1
End of year 2	$1.1 + 1.1 \times 0.10$	1.21
End of year 3	$1.21 + 1.21 \times 0.10$	1.331
End of year 4	$1.331 + 1.331 \times 0.10$	1.464
End of year *n*	$1\,(1 + 0.1)^n$	

It is thus possible to say that £1.464 in 4 years' time at an earnings rate of 10 per cent has a *present value* (*PV*) of £1, or that £1 in 4 years' time at an earnings rate of 10 per cent has a present value of $\frac{1}{1.464}$ = £0.683.

Tables of *PV* factors are readily available for a variety of circumstances (invested immediately, or invested each month and so on). Alternatively pocket calculators with built-in programs can be easily and cheaply obtained.

Projects may have both cash outflows (expenditures) and inflows (payments); the *net present value* (*NPV*) is the sum of *all* cash flows. By comparing the *NPV* of competing projects it is possible to appraise the values of the various projects to the organization.

Discounting

An alternative way of appraising projects is to *discount* the cash flows, discounting being the inverse of compounding. Any project will generate a series of in- and outflows of cash. It is possible to discover the earnings rate that allows the *PV* of the inflows and the outflows to balance. This rate is the *internal rate of return (IRR)*, the *time adjusted return* or the *project rate of return (PRR)*. Discovering this rate allows competing projects to be compared, the most desirable project having the highest rate of return.

The technique is probably best illustrated by an example which, for purpose of clarity is greatly simplified. A project has the following characteristics: annual costs are committed at the beginning of each year, and these are the only costs during the year.

Cost at the beginning of year:	*Value (£)*
1	2,500
2	3,000
3	6,500
4	4,500
5	4,000
Total	20,500

It is anticipated that when the project is complete at the end of year 5 it will generate an income of £32,000. What would be the *PRR*?

The *PRR* is calculated by a trial and error procedure.

Earnings rate 15 per cent

	Cash flow (£)	*PV factor*	*PV(£)*
Beginning of year 1	2,500	1	2,500
Beginning of year 2 (after 1 year)	3,000	0.870	2,610
Beginning of year 3 (after 2 years)	6,500	0.756	4,914
Beginning of year 4 (after 3 years)	4,500	0.658	2,961
Beginning of year 5 (after 4 years)	4,000	0.572	2,288
		NPV	15,273

A cash inflow of £32,000 after 5 years has an *NPV* value of £32,000 × 0.497 = 15,904

Earnings rate 20 per cent

	Cash flow (£)	*PV factor*	*PV(£)*
Beginning of year 1	2,500	1	2,500
Beginning of year 2 (after 1 year)	3,000	0.833	2,499
Beginning of year 3 (after 2 years)	6,500	0.694	4,511
Beginning of year 4 (after 3 years)	4,500	0.579	2,605
Beginning of year 5 (after 4 years)	4,000	0.482	1,928
		NPV	14,043

A cash inflow of £32,000 after 5 years has an *NPV* of £32,000 × 0.402 = 12,864

Earnings rate 25 per cent

	Cash flow (£)	*PV factor*	*PV(£)*
Beginning of year 1	2,500		2,500
Beginning of year 2 (after 1 year)	3,000	0.800	2,400
Beginning of year 3 (after 2 years)	6,500	0.640	4,160
Beginning of year 4 (after 3 years)	4,500	0.512	2,304
Beginning of year 5 (after 4 years)	4,000	0.410	1,640
		NPV	13,004

A cash inflow of £32,000 after 5 years has an *NPV* of £32,000 × 0.328 = 10,496

Given these three sets of figures it is possible to sketch (Fig. 18.6) the *NPV*s of the in- and outflows at various discount factors, the intersection of the two curves indicating the balancing point and hence the *PRR* of, effectively, 17 per cent. Alternatively, a series of 'homing-in' calculations can be carried out. Clearly, attempts at high accuracy are time-consuming and unnecessary.

This value is then used for comparison with other competing projects.

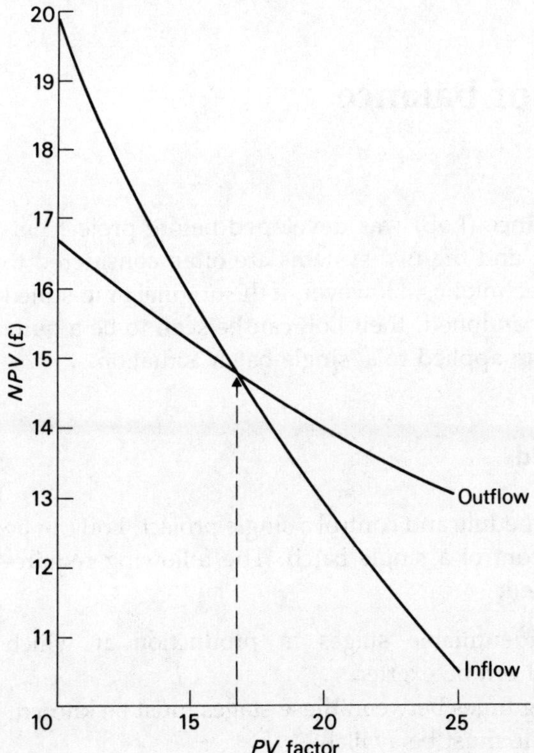

Fig. 18.6 NPVs of cash inflows and outflows at various *PV* factors

Appraisal not costing

It must be emphasized that the above techniques and their variants are methods of judging the desirability of alternative projects—*they do not form the basis of any costing system.*

19 Line of balance

Historically, line of balance (LoB) was developed before project network techniques (PNT), and the two systems are often considered to be separate but related techniques. However, if the original time-scaled stage-time diagram is abandoned, then LoB can be seen to be a quite conventional PNT system applied to a 'single-batch' situation.

Where LoB can be used

Just as PNT is used to schedule and control a single project, LoB can be used to schedule and control a single batch. The following requirements need to be satisfied:

(1) There must be identifiable stages in production at which managerial control can be exerted.
(2) The manufacturing times between these stages must be known.
(3) A delivery schedule must be available.
(4) Resources can be varied as required.

While it is possible to use LoB to control a number of separate batches, just as it is possible to use PNT to control a number of separate projects, the computational difficulties become great. It is therefore usual to employ LoB in 'single batch' situations where the batch concerned is of some considerable importance to the organization. An estate of houses, a batch of guided weapons or a batch of computers are likely to be the type of work appropriate to LoB control.

Again, as with PNT, LoB can be used in a hierarchical manner, considerable detail and a small time span being displayed at the bottom of the hierarchy, while little detail but a considerable time span is shown at the top.

LoB in use

The LoB technique will be illustrated by reference to the following hypothetical example.

Product *Z* is assembled from five components, *A*, *B*, *C*, *D* and *E*. *A* is purchased outright and *B* is made, tested and then joined with *A* to make sub-assembly 1 (S/A1). *C* is also made and tested, and then assembled with S/A1 to give sub-assembly 2 (S/A2). The material for *D* has to be purchased, and it can then be made up and tested, and then joined with S/A2 to give sub-assembly 3 (S/A3). *E* is a purchased item which is assembled to S/A3 at the final assembly stage to give the complete product *Z*. This final assembly stage can be considered to include the act of delivering the product to the customer. The delivery schedule is as follows: first delivery in week ending 1st January:

Week number	*Quantity*
1	2
2	4
3	8
4	12
5	10
6	10
7	16
8	18
9	20
10	22
11	24
12	26
13	28
14	24
15	10
16	6
17	4
18	2
19	2
20	2
Total	250

Step 1

Construct an arrow diagram to show the logic and timing of the production (Fig. 19.1). It will usually be found most convenient to start to

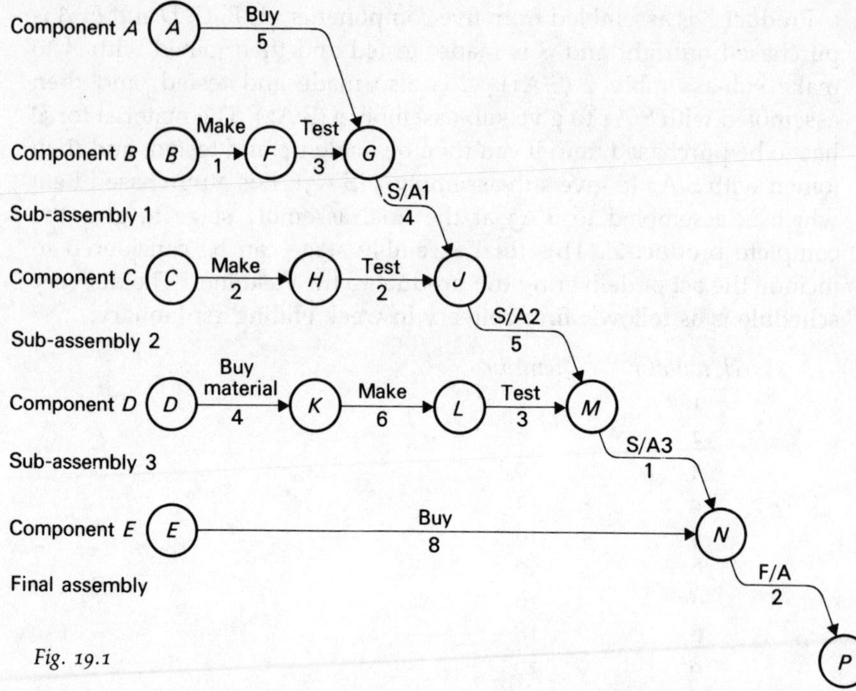

Fig. 19.1

draw this from the end (in this case 'final assembly'), and work towards the various opening activities. The network need not be closed at the start—multiple starts are quite permissible and useful here—and nodes need not necessarily be identified, although for the purposes of the present text the nodes are identified here by letters. Duration times indicate the time required for unit production: these times are maintained constant during production by the variation of resources. The final chart is now very similar to the 'GOZINTO' diagram discussed by—for example—Vaszonyi.

Step 2

Carry out a reverse forward pass from time 0 at the final event, that is, assign to the final node a time 0, and then successively add duration times for each activity in order. This will give the set of figures inscribed against each node, 2 at *N*, 3 at *M*, 8 at *J* and so on (Fig. 19.2).

The result of this reverse forward pass can also be represented on a time-scaled diagram, which is the form in which LoB results are often presented (Fig. 19.3).

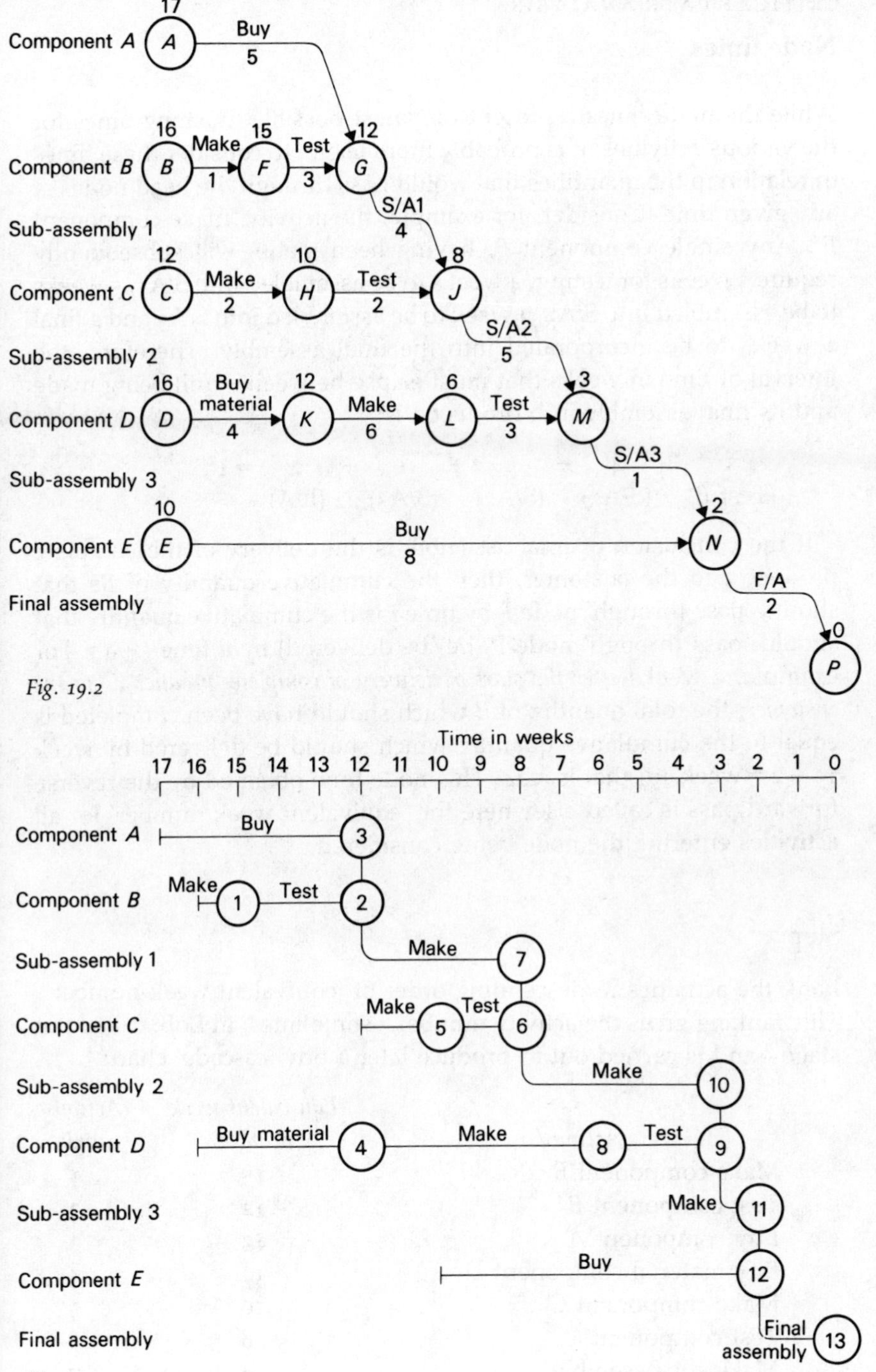

Fig. 19.2

Fig. 19.3

Node times

While the node times represent the latest possible finishing times for the various activities, it is probably more useful to consider these times in relation to the quantities that would pass through the head nodes at any given time. Consider, for example, the activity 'make component *B*'. Any single component *B*, having been made, will subsequently require 3 weeks for testing, 4 weeks to be assembled into S/A1, 5 weeks to be assembled into S/A2, 1 week to be assembled into S/A3 and a final 2 weeks to be incorporated into the final assembly. Therefore, the interval of time in weeks that must elapse between a unit being made and its final assembly into product 2 is:

$$\underset{(\text{Test } B)}{3} + \underset{(\text{S/A1})}{4} + \underset{(\text{S/A2})}{5} + \underset{(\text{S/A3})}{1} + \underset{(\text{F/A})}{2} = 15$$

If the conclusion of final assembly is the delivery of the complete product 2 to the customer, then the cumulative quantity of *B*s that should 'pass through' node *F* by time t is the cumulative quantity that should 'pass through' node *P* (i.e. be delivered) by a time $t+15$. For example, 2 weeks *after the start of delivery of complete 'product Z' to the customer*, the total quantity of *B* which should have been completed is equal to the cumulative quantity which should be delivered by week $15+2=$ week 17, that is, 244. This node time obtained by the reverse forward pass is called elsewhere the 'equivalent week number' for all activities entering the node being considered.

Step 3

Rank the activities in descending order of 'equivalent week number'. This ranking gives the activity number—sometimes, in LoB, called the stage—and is carried out to produce later a tidy 'cascade' chart:

Activity	*Equivalent week number*	*Activity number*
Make component *B*	15	1
Test component *B*	12	2
Buy component *A*	12	3
Buy material component *D*	12	4
Make component *C*	10	5
Test component *C*	8	6
Make sub-assembly 1	8	7

Activity	*Equivalent week number*	*Activity number*
Make component *D*	6	8
Test component *D*	3	9
Make sub-assembly 2	3	10
Make sub-assembly 3	2	11
Buy component *E*	2	12
Carry out final assembly	0	13

Step 4

Prepare a calendar and accumulated delivery quantity table:

Date		*Week number*	*Quantity*	*Cumulative quantity*
4th	September	−17		
11th	September	−16		
18th	September	−15		
25th	September	−14		
2nd	October	−13		
9th	October	−12		
16th	October	−11		
23rd	October	−10		
30th	October	−9		
6th	November	−8		
13th	November	−7		
20th	November	−6		
27th	November	−5		
4th	December	−4		
11th	December	−3		
18th	December	−2		
25th	December	−1		
1st	January	1	2	2
8th	January	2	4	6
15th	January	3	8	14
22nd	January	4	12	26
29th	January	5	10	36
5th	February	6	10	46
12th	February	7	16	62
19th	February	8	18	80
26th	February	9	20	100
5th	March	10	22	122

Date		*Week number*	*Quantity*	*Cumulative quantity*
12th	March	11	24	146
19th	March	12	26	172
26th	March	13	28	200
2nd	April	14	24	224
9th	April	15	10	234
16th	April	16	6	240
23rd	April	17	4	244
30th	April	18	2	246
7th	May	19	2	248
14th	May	20	2	250

Step 5

From the above two tables deduce the quantity of each activity which should be completed by any particular date. For example:

> It is now 22nd January. How many of each component should be completed?
> Consider 'make Component *D*'.
> The time is now week 4.

The quantity through 'make component *D*' is equal to the quantity which can pass through the final stage in 6 weeks time, that is, in week $4+6=10$. From the table in step 4 this is a total of 122 units.

Similarly, for all the activities:

	Volume of work completed is equivalent to volume delivered at week	*Total units*
Make component *B*	$4+15=19$	248
Test component *B*	$4+12=16$	240
Buy component *A*	$4+12=16$	240
Buy material component *D*	$4+12=16$	240
Make component *C*	$4+10=14$	224
Test component *C*	$4+8=12$	172
Make sub-assembly 1	$4+8=12$	172
Make component *D*	$4+6=10$	.122
Test component *D*	$4+3=7$	62
Make sub-assembly 2	$4+3=7$	62
Make sub-assembly 3	$4+2=6$	46
Buy component *E*	$4+2=6$	46
Carry out final assembly	$4+0=4$	26

This can be represented on a chart—the traditional LoB chart (Fig. 19.4).

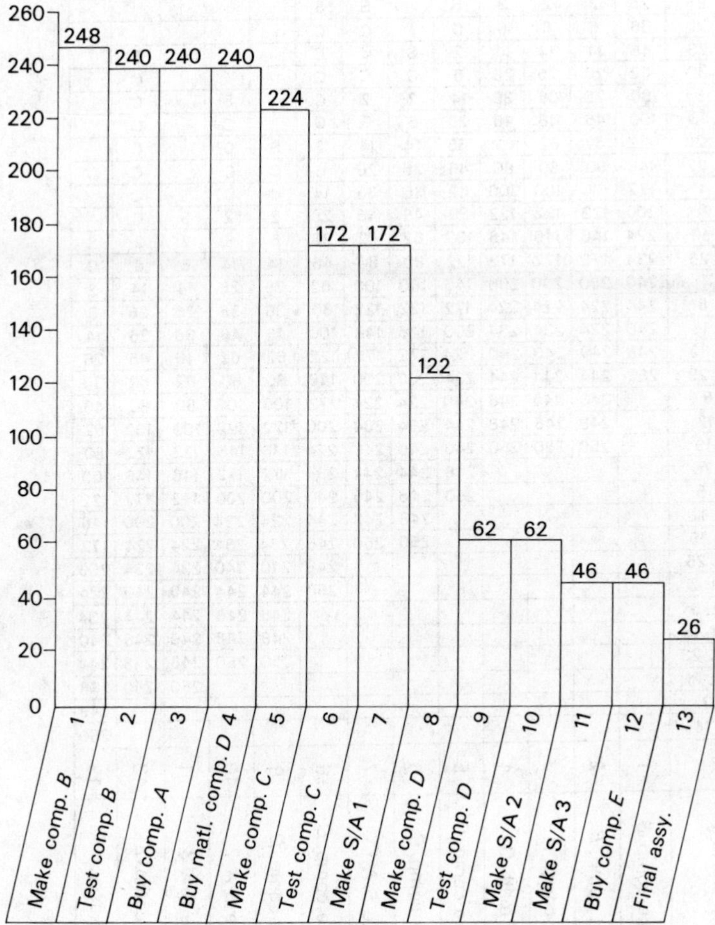

Fig. 19.4

A complete table for the whole 'life' of the batch can be drawn up if desired (Fig. 19.5). The '*S*s' in the table indicate the latest dates by which the various chains of activities should start, this date being derived from the equivalent week numbers from the opening activities. The '*C*s' in the table show that work must be continued.

Week number	Week starting	Make comp. *B*	Test comp. *B*	Buy comp. *A*	Buy matl. *D*	Make comp. *C*	Test comp. *C*	Make S/A 1	Make comp. *D*	Test comp. *D*	Make S/A 2	Make S/A 3	Buy comp. *E*	Final assy.
		1	2	3	4	5	6	7	8	9	10	11	12	13
−17	SEPT. 4			S										
−16	SEPT. 11	S		C	S									
−15	SEPT. 18	2	S	C	C									
−14	SEPT. 25	6	C	C	C									
−13	OCT. 2	14	C	C	C									
−12	OCT. 9	26	2	2	2	S		S	S					
−11	OCT. 16	36	6	6	6	C		C	C					
−10	OCT. 23	46	14	14	14	2	S	C	C				S	
−9	OCT. 30	62	26	26	26	6	C	C	C				C	
−8	NOV. 6	80	36	36	36	14	2	2	C		S		C	
−7	NOV. 13	100	46	46	46	26	6	6	C		C		C	
−6	NOV. 20	122	62	62	62	36	14	14	2	S	C		C	
−5	NOV. 27	146	80	80	80	46	26	26	6	C	C		C	
−4	DEC. 4	172	100	100	100	62	36	36	14	C	C		C	
−3	DEC. 11	200	122	122	122	80	46	46	26	2	2	S	C	
−2	DEC. 18	224	146	146	146	100	62	62	36	6	6	2	2	S
−1	DEC. 25	234	172	172	172	122	80	80	46	14	14	6	6	C
1	JAN. 1	240	200	200	200	146	100	100	62	26	26	14	14	2
2	JAN. 8	244	224	224	224	172	122	122	80	36	36	26	26	6
3	JAN. 15	246	234	234	234	200	146	146	100	46	46	36	36	14
4	JAN. 22	248	240	240	240	224	172	172	122	62	62	46	46	26
5	JAN. 29	250	244	244	244	234	200	200	146	80	80	62	62	36
6	FEB. 5		246	246	246	240	224	224	172	100	100	80	80	46
7	FEB. 12		248	248	248	244	234	234	200	122	122	100	100	62
8	FEB. 19		250	250	250	246	240	240	224	146	146	122	122	80
9	FEB. 26					248	244	244	234	172	172	146	146	100
10	MAR. 5					250	246	246	240	200	200	172	172	122
11	MAR. 12						248	248	244	224	224	200	200	146
12	MAR. 19						250	250	246	234	234	224	224	172
13	MAR. 26								248	240	240	234	234	200
14	APRIL 2								250	244	244	240	240	224
15	APRIL 9									246	246	244	244	234
16	APRIL 16									248	248	246	246	240
17	APRIL 23									250	250	248	248	244
18	APRIL 30											250	250	246
19	MAY 7													248
20	MAY 14													250

Fig. 19.5

Step 6

Record the actual progress upon either the LoB chart or the 'life' table. For example, if at 22nd January the achieved and planned results are:

		Achieved	Planned
1	Make component *B*	200	248
2	Test component *B*	200	240
3	Buy component *A*	200	240
4	Buy material component *D*	200	240
5	Make component *C*	200	224
6	Test component *C*	200	172
7	Make sub-assembly 1	190	172
8	Make component *D*	200	122
9	Test component *D*	200	62
10	Make sub-assembly 2	150	62
11	Make sub-assembly 3	100	46
12	Buy component *E*	90	46
13	Final assembly	90	26

the LoB chart will be as in Fig. 19.6 while the life table will be as in Fig. 19.7.

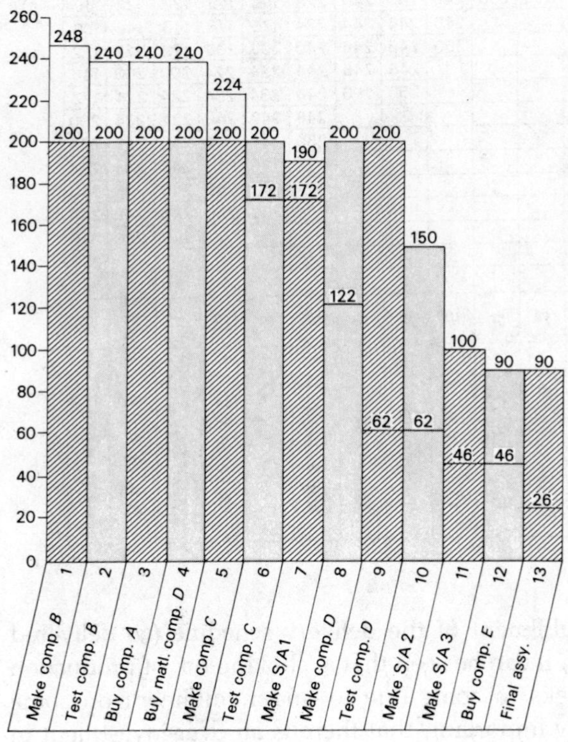

Fig. 19.6

Week number	Week starting	Make comp. *A* 1	Test comp. *B* 2	Buy comp. *A* 3	Buy matl. *D* 4	Make comp. *C* 5	Test comp. *C* 6	Make S/A 1 7	Make comp. *D* 8	Test comp. *D* 9	Make S/A 2 10	Make S/A 3 11	Buy comp. *E* 12	Final assy. 13
−17	SEPT. 4			S										
−16	SEPT. 11	S		C	S									
−15	SEPT. 18	2	S	C	C									
−14	SEPT. 25	6	C	C	C									
−13	OCT. 2	14	C	C	C									
−12	OCT. 9	26	2	2	2	S		S	S					
−11	OCT. 16	36	6	6	6	C		C	C					
−10	OCT. 23	46	14	14	14	2	S	C	C				S	
−9	OCT. 30	62	26	26	26	6	C	C	C				C	
−8	NOV. 6	80	36	36	36	14	2	2	C		S		C	
−7	NOV. 13	100	46	46	46	26	6	6	C		C		C	
−6	NOV. 20	122	62	62	62	36	14	14	2	S	C		C	
−5	NOV. 27	146	80	80	80	46	26	26	6	C	C		C	
−4	DEC. 4	172	100	100	100	62	36	36	14	C	C		C	
−3	DEC. 11	200	122	122	122	80	46	46	26	2	2	S	C	
−2	DEC. 18	224	146	146	146	100	62	62	36	6	6	2	2	S
−1	DEC. 25	234	172	172	172	122	80	80	46	14	14	6	6	C
1	JAN. 1	240	200	200	200	146	100	100	62	26	26	14	14	2
2	JAN. 8	244	224	224	224	172	122	122	80	36	36	26	26	6
3	JAN. 15	246	234	234	234	200	146	146	100	46	46	36	36	14
4	JAN. 22	248	240	240	240	224	172	172	122	62	62	46	46	26
5	JAN. 29	250	244	244	244	234	200	200	146	80	80	62	62	36
6	FEB. 5		246	246	246	240	224	224	172	100	100	80	80	46
7	FEB. 12		248	248	248	244	234	234	200	122	122	100	100	62
8	FEB. 19		250	250	250	246	240	240	224	146	146	122	122	80
9	FEB. 26					248	244	244	234	172	172	146	146	100
10	MAR. 5					250	246	246	240	200	200	172	172	122
11	MAR. 12						248	248	244	224	224	200	200	146
12	MAR. 19						250	250	246	234	234	224	224	172
13	MAR. 26								248	240	240	234	234	200
14	APRIL 2								250	244	244	240	240	224
15	APRIL 9									246	246	244	244	234
16	APRIL 16									248	248	246	246	240
17	APRIL 23									250	250	248	248	244
18	APRIL 30											250	250	246
19	MAY 7													248
20	MAY 14													250

Fig. 19.7

Despite the over-fulfilment of the delivery schedule (90 delivered and only 26 required), it can be seen that a 'choking-off' of production will occur in the weeks to come due to under fulfilment on some activities, and, equally important, that there is an over-investment in work-in-progress on other activities. It may, therefore, be possible to transfer resources from the 'rich' activities to the 'poor' ones while preserving the delivery schedule: decisions here can only be taken in the light of local knowledge, and will require reference to both the PNT diagram and the progress results.

Design/make projects—joint PNT, LoB

It is not uncommon to find projects that involve a setting-up stage (design, plan, make jigs and tools) followed by the production of a batch of equipment. Here it is possible to use conventional PNT for all the work up to, and including the making of the first complete equipment, and then to employ LoB to control the subsequent batch production.

Appendix Using the computer

It would be an act of extreme folly for a project planner to write a PNT computer program since many are readily available. As representative of these, three programs from:

K&H Business Consultants Ltd.
9 Villiers Road
Kingston-on-Thames
Surrey, England KT1 3AP

are outlined here. This does not imply any recommendation from the author that these are the 'best' or the 'ideal', but that they represent good modern practice:

(1) G/C Cue
(2) Amper/Premis (precedence version)
(3) Amper/Premis

G/C Cue

G/C Cue is a project control system developed for the Hewlett-Packard 3000 and for the PRIME 350 minicomputers. It provides on-site control that is flexible, interactive and user-oriented, and it consists of seven sub-systems:

(1) *Cost and performance measurement*: integrates accounting, estimating and scheduling information to measure performance, revise estimates, prepare cash and quantity flows, and determine project status by earned value.
(2) *Construction accounting*: records and reports all actual and committed costs.
(3) *Cost estimating*: prepares, records and updates all project estimates.
(4) *Planning and scheduling*: continuously monitors the effect of changes in task completion dates.
(5) *Data management*: supports both document and quantity tracking, and organizes and lists the large volumes of data that are part of major projects.

(6) *Property records*: an optional sub-system that allocates direct and indirect expenses according to corporate or governmental requirements.
(7) *Payroll*: this optional sub-system provides responsive on-line processing of hourly and salaried personnel payrolls.

Codes are used to build numbers, designate, sort and select criteria, and introduce special variables according to the user's own specification. Numbers are used for identifying cost accounts, planning and scheduling network activities, and for identifying such data as purchase orders and invoices. It uses the work breakdown structure to organize and store information.

Amper/Premis (precedence version)

Activities in this system are identified by an alphanumeric code of up to 18 characters. The program will handle sub-projects of up to 64,000 activities. Each activity may be dependent on up to 100 other activities (depending on the length of the activity number).

The relation or dependency between activities may be:

(1) Start-to-start (*S*).
(2) Finish-to-finish (*F*).
(3) Finish-to-start (*N*).
(4) Parallel (*P*).
(5) Start-to-finish (*E*).

and may have durations (lags) associated with them.

Resources are identified by a three-character code. Resource requirements may be specified on an activity as either total work content or as a rate over every time unit. Complex resource patterns may be specified over the duration of the activity.

There may be up to one hundred resource entries on any activity; resource, material and all types of costs may be associated with activities.

Given the relevant rates, the program will calculate total resource costs on activities.

Amper/Premis

Amper is a critical path/resource analysis system and is able to handle sub-nets of up to 80,000 activities. Sub-nets may be grouped into networks, which may be grouped into sub-systems. Sub-nets within the same sub-system may be interfaced through common nodes.

Resources are identified by a three-character code, and may be specified on the activity in three different ways:

(1) Rate constant.
(2) Total constant.
(3) Work content.

Complex resource requirement patterns may be specified over the duration of the activity (up to one hundred different patterns per activity).

There may be up to one hundred resources on any activity. Activities may appear in more than one place in a report if required (e.g. sorts by resource used, multiple areas, etc.).

There is a large range of possibilities in updating and progress can be reported in many different ways. Extraneous information of almost any type may be associated with activities and used in reports and calculations. Historical data of varying kinds may be retained for comparison and reporting purposes.

The input format is entirely at the discretion of the user—there can be any number of cards and/or card types.

Questions

All the questions follow broadly the same form namely some text indicates the procedure, followed by a list of activities. In most, but not all, duration times are provided.

It is suggested that after reading the chapters concerned with the drawing of networks the reader should sketch the network implicit in each question, ignoring duration times. The logic of the sketch should then be checked against the model answer—a solutions manual is obtainable on demand from the publishers.

When the reader is satisfied that he/she can draw a network, the chapters on analyzing a network should be studied, and when understood, duration times applied to the sketches and an analysis carried out. This analysis can then be checked against the model answer. *Note: Use a pencil on large sheets of paper, and keep an eraser handy, when drawing networks.*

The author has found that whenever a difficulty in analysis arises, it can *invariably* be resolved by drawing a bar (Gantt) chart. Even if the question does not ask for a bar chart it will be found rewarding to draw one.

No general resource allocation problems are provided. This is for two reasons: firstly, decision rule techniques—the most commonly used methods of resource allocation—do not necessarily provide unique answers, and therefore the student's answer may differ from the author's, yet both are equally valid. Secondly, manual resource allocation is tedious in the extreme and very few students have the persistence to 'follow through' a resource problem. This having been said, students interested in this topic are recommended to apply arbitrarily resources to problems, and draw the resultant histograms in the 'earliest start' and 'latest start' positions.

Questions 19 and 20 are specifically intended to illustrate precedence networking. Question 19 is capable of being solved by any system of networking: question 20 is a precedence network containing all relationships for analysis.

Question 1: four parts from three machines

In the manufacture of a piece of apparatus, the final assembly (which has to be tested by a piece of specially made test gear) is made up from two items:

(1) Component *D*.
(2) A second assembly of parts.

The second assembly in turn is made up from two items:

(1) Component C.
(2) A first assembly.

and the first assembly is made up from two parts:

(1) Component *A*.
(2) Component *B*.

To manufacture these components special machines must be obtained as follows:

First machine	produces component *A*
Second machine	produces component *B*
Third machine	produces components C and *D*

Each component is tested before it is assembled with another component or item, but test gear can be assumed to be available for all testing except final testing where special test gear has to be made. The design of this special test gear is known at the outset of the whole project. The activities involved in the project are:

Activities involved	*Duration (days)*
Obtain first machine	2
Obtain second machine	3
Obtain third machine	2
Make component *A*	2
Make component *B*	2
Make component *C*	3
Make final test gear	20
Test component *A*	3
Test component *B*	4
Test component *C*	3
Make first assembly	3

Activities involved	*Duration (days)*
Test final assembly	5
Make component *D*	8
Test component *D*	20
Make second assembly	7
Make final assembly	3

Questions

(1.1) Prepare a schedule showing the times when the various activities must be carried out in order that the total project can be completed in the minimum time.

(1.2) Draw a bar and/or Gantt chart for the project.

Question 2: the ATC tower

An air traffic control (ATC) tower utilizes an ATC console which, though standard in terms of mechanical design and input/output panels, has to be electrically designed to meet the needs of the particular airfield. The activities involved in building and equipping an ATC tower are:

Activities involved	*Duration (weeks)*
Design console	3
Order console	2
Make and deliver console	16
Install console	2
Operationally test console	4
Design tower	5
Design foundations	1
Order foundations material	1
Deliver foundations material	1
Construct foundations	1
Order tower constructional material	2
Deliver tower constructional material	3
Erect tower	8

Questions

(2.1) What is the minimum time that must elapse between the

receipt of a 'letter of intent' (i.e. a purchaser order) for a fully equipped tower and the 'hand-over' date (i.e. the tower being fully equipped with tested material)?

(2.2) What would be the effect of a delay in the designing of the tower of:

(*a*) 1 week?
(*b*) 2 weeks?
(*c*) 3 weeks?
(*d*) 5 weeks?

(2.3) What action, if any, appears desirable if 10 weeks after receipt of the letter of intent the following activities are complete:

Design console	Design tower
Order console	Order tower constructional material

and the following activities are started but not yet complete:

'Make and deliver console' requiring 11 weeks more work.
'Obtain tower constructional material' requiring 2 weeks more work.

and no other activities are yet started?

(2.4) What saving in time, if any, would have resulted if a completely standard console had been used and the tower designed to match it so that the following times had been obtained:

Design console	1 week	Make and deliver console	8 weeks
Order console	2 weeks	Design tower	8 weeks

all other times remaining as in the table above?

Question 3: the telescopic gun-sight

(An exercise suggested by a publication of the Industrial Operations Unit of the Department of Scientific and Industrial Research entitled 'Application of Critical Path Method of Scheduling: A Demonstration'. The modifications to the DSIR report are so great that the exercise can now be considered to be completely fictitious.)

A manufacturer has a telescopic gun-sight assembly that he makes up only upon receipt of a firm order from a customer. The design of this gun-sight is completely stable, and no deviations from the standard product are ever accepted, with the result that all drawings, tools and production aids are always available.

The parts used in the assembly can be considered to be as follows:

(1) Tube.
(2) Eye-piece.
(3) Lens holder.
(4) Locknut.
(5) Lens nut.
(6) Washer.
(7) Spring.
(8) Screw.
(9) Lens.

Parts 1–5 were manufactured internally after purchase of the initial raw material (aluminium) and were sent outside for finishing (anodizing). Parts 6–9 were bought complete, ready for assembly, from outside suppliers.

Upon receipt of a firm order, an internal works order is raised, which initiates all activity. The relevant drawings are then extracted from the drawing library and copies sent to the appropriate departments. Having received the drawings, the material control section scans its records and raises purchase requisitions, which are forwarded to the purchasing department. Purchase orders are then made out and once materials are received they are inspected and passed either to the machine shop (in the case of raw material) or to the final assembly department. Once all parts are available in the final assembly department, they are assembled into the complete gun-sight, which is then inspected, packed and eventually dispatched to the customer. A list of operations follows:

Operation	*Duration (days)*
Raise works order	2
Extract and circulate drawings	4
Raise purchase requisitions	3
Prepare production paperwork	5
Issue production paperwork	2
Place purchase orders	4
Piece part tools withdrawn from store	1
Final assembly tools withdrawn from store	2
Raw material delivered	4
Part 6 delivered	22
Part 7 delivered	24
Part 8 delivered	10
Part 9 delivered	30
Part 1 machined	4
Part 2 machined	8
Part 3 machined	12

Operation	*Duration (days)*
Part 4 machined	7
Part 5 machined	3
Inspect part 1 before finishing	2
Inspect part 1 after finishing	1
Inspect part 2 before finishing	2
Inspect part 2 after finishing	1
Inspect part 3 before finishing	3
Inspect part 3 after finishing	1
Inspect part 4 before finishing	4
Inspect part 4 after finishing	2
Inspect part 5 before finishing	2
Inspect part 5 after finishing	1
Inspect part 6	1
Inspect part 7	1
Inspect part 8	1
Inspect part 9	1
Anodize part 1	5
Anodize part 2	5
Anodize part 3	5
Anodize part 4	5
Anodize part 5	5
Raw material inspected	1
Final assembly	10
Final inspection	4
Pack	4
Dispatch	2

Questions

(3.1) A firm order is received on Monday, 30th December, and work is immediately put in train.

(i) What is the earliest date by which delivery can be completed?

(ii) Past experience has shown that the inspection in the machine shop has proved a bottleneck. What could be done to allow one inspector to do all the machine shop inspection without delaying the completion of the job?

(iii) Assume that parts 1 and 2 are made in machine *A*, parts 4 and 5 on machine *B*, and that only one machine *A* and one machine *B* are available, what could be done to allow one inspector to do all the machine shop inspection without delaying the completion of the job? (Assume no limitation on availability of machine to make part 3.)

(3.2) Assuming that there are no resource limitations and no changes in duration time and/or logic are made in the manufacture of the gun-sight, examine the situation that would exist if, 20 days after the receipt of the customer's order, the following operations had been completed as early as possible:

Raise works order.
Extract and circulate drawings.
Prepare production paperwork.
Issue production paperwork.
Issue piece part tools.
Issue final assembly tools.
Raise purchase requisitions.
Place purchase orders.

and it was understood that the time for delivery of raw materials would be increased from 4 days to 14 days, and that delivery of part 7 would increase from 24 to 26 days. All other operations were expected to be performed in the times originally estimated.

(3.3) Believing that the overall time from receipt of the customer's order to dispatching the finished goods was too great, the general manager invited suggestions for reducing this time. He specified that the cost was not an important consideration in this case, and that capital expenditure could be tolerated providing that reduction in total time could be demonstrated. He received the following suggestions:

(i) *From the material control section*: hold bigger stocks of raw material so that machining could start by withdrawing material from stores rather than obtaining it from an outside supplier.

(ii) *From the production manager*: purchase a piece of equipment for £4,000 that would enable part 6 to be made internally in 4 days and another piece of equipment for £3,000 that would enable part 7 to be made internally in 5 days. Inspection time would not be reduced on either part.

(iii) *From the office manager*: purchase a more efficient office duplicator that would enable all the works orders to be raised in 1 day. The cost of this equipment would be of the order of £350.

(iv) *From the chief inspector*: abandon the pre-finishing inspection for the machined parts (parts 1–5) but increase the scope of the post-finishing inspection. This would double the time for the post-finishing inspection and increase the risk that defective parts were finished. The chief inspector stated that this risk was very small.

(v) *From the foreman of machine shop 3*: permit overtime working for machine shop 3 in order to reduce the machining time from 12 working days to 10 working days.

(vi) *From the chief draughtsman*: purchase an electrically operated printing machine to enable the time for the extraction of drawings to be reduced from 4 days to 3 days. Cost estimated to be £450.

(vii) *From the production controller*: purchase a new duplicating machine (cost £100) to reduce the time for raising production paperwork by one half.

(viii) *From the buyer*: part 9 can be obtained from a different source at an increased cost but with a delivery time of 20 days.

Evaluate the above suggestions.

Question 4: nine phrases

Sketch the following nine 'phrases'. Since they form parts of complete networks, they are not necessarily complete in themselves, that is, they do not start from, nor do they finish on, a single event.

(i) Task K depends on tasks A and B.

(ii) Task K and task L depend on tasks A and B.

(iii) Task K depends on tasks A and B and task L depends only on task B.

(iv) Task K depends only on task A, but task L depends on both tasks A and B.

(v) Task K depends on tasks A and C, and task L depends on tasks B and C.

(vi) Task K depends on tasks A and C, task L depends on tasks B and C and task M depends only on task C.

(vii) Task K depends on task A, task L depends on task B, and task M depends on tasks A, B and C.

(viii) Task K depends on task A, task L depends on tasks A and B, and task M depends on tasks B and C.

(ix) Task K depends on task A, task L depends on tasks A and B, and task M depends on tasks A, B and C.

Question 5: excavate–shutter–pour

Foundations are to be excavated, and shuttering erected to receive poured concrete. The three activities take the following times:

Activities involved	*Duration (days)*
Excavate foundations	24
Erect shuttering	12
Pour concrete	18

If the three activities are carried out in sequence, each activity being completed before its successor is started, the total time for the project will be 54 days. To reduce this it is decided to start shuttering when only part of the excavating is complete, and pouring when only part of the shuttering is complete. Draw the arrow diagrams for the situations when work on a successor starts:

(i) After one-half
(ii) After one-third
(iii) After one-quarter

of a predecessor is complete. Assume that only one excavating gang, one shuttering gang and one pouring gang are available, and that these gangs must not be split. Calculate the total times for (i), (ii) and (iii) and then express them as a percentage of the original total time of 54 days.

Question 6: draw–trace–print

In the preparation of a set of drawings the following activities occur:

Activities involved	*Duration (weeks)*
Draw	12
Trace	8
Print	4

The chief draughtsman is prepared to release drawings to the tracers after 2 weeks, and the tracers will release drawings to the print room after 1 week. Sketch the network for this situation.

Question 7: network V

Draw the network represented by:

	Activity	*Precedes*	*Duration*
START	*A*	*D, E*	4
START	*B*	*F, G*	6
START	*C*	*G*	2
	D	*L*	3
	E	*H*	8
	F	*H, K*	9
	G	*L*	10
	H	*L*	6
	K	*L*	8
FINISH	*L*	—	1

Analyze this network and draw a bar chart.

Question 8: overhauling delivery lorries

A fleet of three delivery lorries are overhauled each week. The practice is that Joe, the sole mechanic, overhauls a lorry, and then Fred, the sole garage-hand, cleans and polishes that lorry.

There are three lorries, *A*, *B* and *C*, and they are processed in that order.

Sketch the network for the above situation. Use the following activities:

Overhaul *A*	Clean and polish *A*
Overhaul *B*	Clean and polish *B*
Overhaul *C*	Clean and polish *C*

Question 9: a new machine

Draw the network for the following:

A new machine is required for which budget approval is required. The use of the new machine necessitates employing a new operator who must be specially trained, using a training manual and some special equipment that is delivered with the machine. The training itself does not depend upon the new machine being installed and working. Once the machine is installed and the operator is trained, production can proceed.

Activities involved

Obtain budget approval
Obtain machine
Install machine
Hire operator
Train operator
Production of goods

Question 10: the machine tool overhaul

Draw the network for the following situation:

Within a factory there are three machine tools that have to be removed from their bases, modified and then reinstalled on the same bases. Assume that there is only one heavy gang to undertake the initial moving, only one fitter capable of carrying out the modifications, and only one gang of millwrights capable of reinstalling the machines.

Activities involved

Remove first machine
Remove second machine
Remove third machine
Modify first machine
Modify second machine
Modify third machine
Install first machine
Install second machine
Install third machine

Question 11: 'the rubble fill . . .'

(Courtesy of Building Research Station)

Draw the network for the following situation:

The rubble fill cannot be started until the brickwork below d.p.c., the drainage under the building and the mainlaying under the building are finished. The completion of the drainage will follow after the drainage under the building and the external mainlaying will follow the mainlaying under the building. The external paving must be laid after the external mainlaying and the brickwork below d.p.c. The completion of the drainage and the external mainlaying must be finished before the internal services.

Activities involved

Start internal services
Finish drainage under the building
Finish brickwork below d.p.c.
Finish mainlaying under the building
Complete drains
Complete external mainlaying
Lay external paving
Install internal services
Fill with rubble

Question 12: the stock-taking

Within a small neighbourhood department store there are three sections in the shoe department:

(1) Men's.
(2) Women's.
(3) Children's.

There are three staff:

(1) A junior.
(2) A management trainee.
(3) A manager.

The manager has to have a stock-take and decides to do it as follows:

(1) The junior will remove stock, clean fixtures thoroughly and replace stock conveniently for stock-taking, not being considered experienced enough to do more than this.
(2) The management trainee will then count and record the stock.
(3) The manager will sample check the trainee's stock-take.

It is decided to carry out the work section by section, starting with the men's section, following with the women's and finishing with the children's section.

Draw the network for the above, considering only the following activities:

Remove and clean men's stock	R_M
Remove and clean women's stock	R_W
Remove and clean children's stock	R_C
Carry out men's stock-check	S_M
Carry out women's stock-check	S_W
Carry out children's stock-check	S_C
Check men's stock-take	C_M
Check women's stock-take	C_W
Check children's stock-take	C_C

Question 13: network W

The logical relationships and duration times for the activities of a project are:

	Activity	*Precedes activities*	*Duration*
Opening	*A*	*D, E, F*	5
	B	*G*	4
	C	*H*	6
	D	*G*	7
	E	*H*	8
Closing	*F*	—	10
	G	—	11
	H	—	10

Draw the network for the above and analyze it.

Question 14: network X

The logical relationships and duration times for the activities of a project are:

	Activity	Precedes activities	Duration
Opening	*A*	*D, E*	4
	B	*E*	3
	C	*G*	5
	D	*F, G*	7
	E	*H, J, K*	2
	F	*N*	2
	G	*L*	6
	H	*L*	4
	J	*M*	9
	K	—	12
Closing	*L*	—	4
	M	—	8
	N	—	3

Draw the network for the above and analyze it.

Question 15: network Y

The logical relationships and duration times for the activities of a project area.

	Activity	Precedes	Duration
Opening	*A*	*H, L*	43
	B	*J, L*	40
	C	*K, L*	16
	D	*L, M*	37
	E	*L, N*	30
	X	*T, Y*	20
Across whole project	*Z*	—	165
	H	*P*	30
	J	*Q*	29
	K	*R*	40
	L	*P, Q, R, S, T*	2
	M	*S*	70
	N	*T*	62
Close	*P*	—	45
	Q	—	60
	R	—	70
	S	—	35
	T	—	45
	Y	—	30

Draw and analyze the network.

Question 16: the pre-production models

In the pre-production stages of the manufacturing of a piece of equipment that contains both electrical and mechanical parts, the following procedure is carried out:

The specification is agreed, after which the design department makes a prototype for approval, by the sales department. This prototype is in a form final enough for the later detailed mechanical and electrical design work to proceed independently. Once the detailed design work is complete, further sets of both electrical and mechanical parts are made which are then assembled together to form a series of complete pre-production models (PPMs).

Note: the PPMs require both electrical and mechanical parts.

Enough PPMs are made for electrical and mechanical proving tests to be carried out independently, although both proving tests require special test gear to be made. The manufacture of this test gear can be undertaken once the detailed development is complete. Electrical and mechanical proving tests having been completed, a final approval is given that enables materials to be ordered and production equipment to be made. Once all the material and production equipment is to hand, manufacture can start.

Activities involved	*Duration (days)*
Agree specification	6
Make prototype	12
Prototype approved	2
Detailed mechanical design	12
Detailed electrical design	10
Make mechanical parts for PPMs	6
Make electrical parts for PPMs	6
Make mechanical test gear	15
Make electrical test gear	10
Make pre-production models (PPMs)	12
Electrical proving test PPMs	10
Mechanical proving test PPMs	8
Final approval given	4
Make production equipment	15
Obtain production material	4
Manufacture	15

Questions

(16.1) Draw the network.
(16.2) Analyze the network.
(16.3) Translate the network into a Gantt or other bar chart.
(16.4) List those activities which, 35 days after work has started must:
  (*a*) have been finished
  (*b*) have been started
if the total project time is not to be increased.

Question 17: the storage heater

A Derbyshire fireclay manufacturer has obtained the manufacturing and selling rights of a new storage heater that will utilize some of his own products. It has been decided that initially the heater will be sold only in the London area, since this represents the largest concentrated market, and transport to London will be by rail since the company has its own railway sidings, but the actual volume of heaters to be offered will only be determined after a market survey. This survey will not require the presence of any sample heaters. In order to obtain the rights to the new heater, the Derbyshire manufacturer undertook to market the heater, but the market price was not fixed by the German company, and had to be determined by local conditions.

Manufacture of first batch

Although the design is complete, a number of models will need to be made for test and approval by both the sales and manufacturing departments. Only after these approvals have been obtained, the market survey has been carried out, and a decision on volume has been made, will the production equipment and material be obtained. The heater itself, though inexpensive, is fragile, and special 'immediate' packing will need to be designed, this packing later forming part of the retailer's point-of-sales display. Within this packing will be an 'operating and installation' leaflet, which can only be prepared after the approval of the samples. The design of the packing itself can start once samples are available, but it need not wait upon the testing and approval by either the sales or the production departments since any modifications to the heater as a result of sales and manufacturing approval are not likely to affect the design of the packing. The cost of packing will be insignificant in relation to the cost of the heater. The production costs can only be extracted for pricing purposes after the first batch has been made.

Transport

Transport from the manufacturer's factory sidings in Derbyshire to the retail shops in London will be provided by British Railways, who will provide special containers into which numbers of the packed heaters can be loaded at the factory. These filled containers will be lifted from the factory sidings on to the trains, and thence taken to British Railways' wharves at a suitable North London depot. From these wharves British Railways' vans will distribute to the London shops. It is expected that both the sidings and the wharves will need to be modified in order that the goods can be handled most effectively.

A substantial part of the cost of modifying the British Railways' wharves will have to be borne by the manufacturer, and thus the *total* transport costs can be assumed to consist of two parts:

(1) The costs of modifying the wharves.
(2) The costs of moving the heaters from the works to the shops (the 'handling' costs).

Any charges incurred in modifying the works sidings will be capitalized by the company and not directly incorporated in the selling price of the heater.

As there are a number of apparently suitable wharves in North London, and a number of alternative routes, a test run will be carried out. This will check:

(1) The rail route from Derbyshire to London.
(2) The suitability of the proposed wharves.
(3) The suitability of the British Railways' container.

Before carrying out the test-run, therefore, British Railways need to survey the wharves and obtain a suitable sample container into which sample heaters can be loaded. For this purpose, only temporary immediate packing, of a type readily available, is necessary, and neither sidings nor wharves need to be modified, the containers being 'man-handled' on and off the train. However, once the wharves have been modified it will be necessary to train the wharf staff in their operation. No such training is needed for the factory staff who will work at the factory sidings.

Marketing

Once the sales department has approved the sample heaters, and the volume has been determined by the market survey, the sales campaign can be prepared. No further sales activity can be undertaken, however, until the campaign and its budget have been approved. This approval will include a statement of the marketing cost, and this, together with the production and transport costs, can be used in the derivation of the final selling price. The final agreement of sales outlets depends upon the setting of the selling price. Sales staff are trained after the sales campaign has been agreed, and their training will involve some of the sample heaters. Once trained they can take orders for heaters from the shops (that is, they can 'sell-in'). Distribution of the heaters from the North London wharves can follow thereafter.

Advertising will take the form of 'photograph + text' advertisements in magazines and the advertising can only be finalized when the selling price, the artwork and the text are all available. No advertising will take place until all distribution from wharves to shops has been completed.

Pricing

The final selling price will be determined after:

(1) The total transport costs are agreed.
(2) The cost of producing the first batch is known.
(3) The sales budget is approved.

Activities involved	*Duration (days)*
Carry out market survey	15
Produce samples sufficient for all tests	15
Arrange and hold first meeting with British Railways	4
Survey wharves	14
Obtain estimate for modification to wharves	7
Agree estimate for modification to wharves	7
Modify wharves	15
British Railways obtain sample container	7
Arrange test run from works to wharves	3
Carry out test run	5

Activities involved	Duration (days)
Agree price with British Railways for handling from works to shops (handling costs)	15
Sign contract for modifying wharves	3
Agree total transport costs	5
Obtain railway containers	15
Modify sidings for easy loading from the works to train	10
Train wharf staff	3
Production departments agree sample heater	7
Sales departments agree sample heater	7
Design packing	15
Obtain quotation for packing	5
Sales department approve packing	7
Production department approve packing	3
Obtain packing	10
Obtain production materials and equipment	14
Produce first batch	14
Agree selling price	5
Pack first batch	3
British Railways collect first batch from works and send to wharves	3
Agree retail outlets	15
Collect from wharves and send to shops	3
Advertise 'In your shops now'	10
Extract production costs	3
Prepare sales campaign	21
Agree sales campaign and budget	3
Train sales staff	15
Take orders from shops ('sell-in')	5
Prepare advertising artwork	10
Prepare advertising text	10
Book advertising space	5
Finalize advertising	3
Send sample heaters to British Railways for test run	12
Prepare operating and installation instructions	14

Questions

(17.1) What is the shortest possible time that can elapse between the starting of the whole project (that is, the decision to market the heaters in London) and its completion (that is, advertised heaters being available in the shops)?

(17.2) Once started, it is found that the market survey will take at least 21 days to be completed. What effect will this have?

(17.3) The buyer reports that for an increased price he can obtain packing in 5 days rather than the 10 he had previously specified and as time is essential he requests that he be allowed to purchase the more expensive packing. What decision would you recommend the manufacturer to take?

(17.4) The chairman of the committee that is to meet to agree the total transport costs catches influenza and the meeting of the committee is postponed for 5 days. What is the effect of this?

(17.5) By working over two weekends the production manager manages to make the first batch in 10 days. By how much will the total time for the whole project be reduced?

(17.6) Owing to pressure of work, the chief of the railways' supplies department finds it difficult (though not impossible) to obtain the special containers needed for the heaters in the 15 days he had originally quoted. What increase in time, if any, can be permitted?

(17.7) What activities must be completed 40 working days after the start of the whole project?

(17.8) Assuming that it is now 50 days after 'project-start', you learn that the copy writer preparing the operating and installation instructions will not complete his task for another 20 days. What is the effect of this?

(17.9) Draw a bar chart representing the activities of the sales department once the sales budget has been approved, assuming that the advertising 'In your shops now' is completed by the end of Monday, 3rd October.

Activities involved

Agree selling price
Agree sales outlets
'Sell in'
Collection from wharves and send to shops
Advertise 'In your shops now'
Train sales staff
Prepare advertising artwork
Finalize advertising
Prepare advertising text
Book advertising space

Note: The diagram for this question is particularly involved. It is suggested that readers divide the activities into their responsibility areas—*transport* activities, *general and manufacturing* activities and *marketing* activities—and then draw three separate networks corresponding to these three areas, indicating (for example, by means of double nodes or double arrows) which events or activities are common to two or more networks. Once these three networks have been drawn they can be amalgamated into one final network. This is known as *interfacing* and corresponds to the real-life situation where separate departments (in this case the transport, factory and marketing departments) draw diagrams independently and these are then amalgamated by the general manager.

Question 18: the petrol station

A site with appropriate planning permission, has been obtained for a small petrol station. This will consist, essentially, of two islands, one (the sales island) carrying the cash office and the pumps, and the other (the office island) carrying the manager's office, public toilets and an air compressor. Adjacent to the sales island will be a concrete pit which will house the storage tanks for the petrol (Fig. Q18). The pit and the rest of the site (with the exception, of course, of the two islands) is covered with a concrete hardstanding slab and surrounded by a low perimeter wall. Specific planning and design points are detailed below.

Fig. Q18 The petrol station

Services

Adequate main water and main drainage services are available immediately adjacent to the site so that no special provision must be made for these beyond laying them to such locations on the site as appropriate. There is no electricity on or near the site and the Electricity Board has undertaken to install an electrical supply in the form of a transformer located in a corner of the site.

Petrol pumps

These need to be obtained from the manufacturers, and after being erected on the sales island they are connected to the storage tanks and the power supply. Before use they must be inspected by the local authority to ensure safety and compliance with regulations.

Storage tanks

These are housed in concrete pits and covered by the hard-standing slab. Before they are covered however, the tanks and the associated pipe-work have to be inspected by the local authority.

Manager's office

The manager's office furnishings will include a safe, for the custody of overnight takings. The insurance company insists that the office is efficiently burglar-alarmed; once these alarms are installed the insurance company will inspect them. The alarms are powered by main electricity. All furnishings for the office and toilets must be specially obtained and when delivered must be stored under cover to avoid weather damage. Some furnishings are 'built in' and therefore require painting when in position.

'Free air' supply

Compressed air for inflating tyres will be supplied by an electrically driven compressor, the receiver of which must be inspected by a competent person before the compressor is put into use. The air lines to the 'free air' points are installed with the general underground services, and the points themselves are mounted on the perimeter wall.

Approach road signs

To advertise the petrol station, signs will be erected on the approach roads. Sites for these signs have not been negotiated in detail, although some survey work has been done, but actual site negotiation will not start until the project as a whole is started. It is expected that the signs will be in position by the time that the petrol station is ready for use by the public.

Possession and release of site

Before any work can start on the site, a mobile hut is erected to store tools, furnishings and any weatherprone parts, and to act as the site office. The installation of this hut, along with any other preparatory work is known as 'taking possession'. Similarly, when work on the site is complete, the hut, which has been moved as appropriate, is removed, and all scaffolding and plant taken away. This is known as 'cleaning up the site'.

Activities involved	*Duration (weeks)*
Excavate for sales island base	1
Construct sales island base	1
Construct cash office	2
Obtain pumps	10
Erect pumps	1
Connect pumps	2
Inspector approves pump installation	2
Obtain office furnishings	8
Paint and furnish office and toilets	2
Connect office and toilet lighting	1
Excavate for office island	1
Construct office island base	1
Build offices and toilets including all services	2

Activities involved	*Duration (weeks)*
Install burglar alarm	1
Connect up burglar alarm	1
Insurance company inspects burglar alarm	2
Electricity board installs transformer	10
Connect mains cable to transformer	1
Install area lighting	4
Take possession	1
Set-out and level site	1
Excavate for, and lay, all underground services	1
Excavate for pipe-work and tanks	1
Construct concrete pit	3
Install pipe-work and tanks	2
Obtain pipe-work and tanks	8
Obtain compressor	12
Install compressor	1
Connect power to compressor	1
Competent person tests compressor	2
Backfill and cover tanks	1
Construct concrete hardstanding	2
Construct perimeter wall including air points	2
Connect up air points	1
Clean up site	1
Obtain approach road signs	4
Negotiate sites for approach road signs	12
Erect approach road signs	1
Inspector approves pipe-work and tanks	2

Questions

(18.1) Draw and analyze the network that represents the construction of the petrol station.
(18.2) Of the activities listed, four involve excavation. Assuming that only one mechanical excavator was available, suggest a possible sequence in which the four excavation activities can be performed.
(18.3) List those activities that must be completed by the 12th week after the start of the whole task if the total time for the task is not to exceed 20 weeks.
(18.4) What is the effect of a delay of:

(*a*) 1 week
(*b*) 2 weeks
(*c*) 3 weeks

in 'taking possession' of the site?

(18.5) The contractor proposes to use one gang to carry out the work of 'construct sales island base,' 'construct office island base' and 'construct concrete hard-standing'. In order that the gang concerned is kept fully employed and on site for the shortest possible time, when should the above activities be started?

Question 19: the new CPU

(Derived from IBM's 'System/360 Project Control System: Application Description' H20-0222-0—modified version of example of precedence diagramming and used with their permission.)

It is decided to develop a new CPU, and in discussion at a preliminary planning meeting it is agreed that the following activities are involved in preparing a firm plan for approval by the CPU manager involved.

Activities involved	*Duration (days)*
Develop initial plan	8
Prepare initial data	3
Key-punch initial data	1.5
Initial computer run	4.0
Refine plan as a result of initial computer run	4.0
Prepare revised data derived from refined plan	1.0
Key-punch revised data	0.5
Revised computer run	4.0
Derive firm plan as a result of revised computer run	2.0
CPU manager agrees firm plan	2.0

If these activities were carried out strictly 'end-to-end' the time required would be 30 days, and it was felt that this was excessive, and a second preliminary planning meeting was convened. At this it was agreed that the total time for the project could be reduced by starting some activities before their predecessors were complete:

(1) Initial data preparation could start 3 days after the start of the initial plan development, these two activities finishing at substantially the same time.

(2) Key-punching initial data could start after 1 day of initial data preparation, but it would require at least half a day after the finish of data preparation for its completion.
(3) Revised data preparation could start when half the refined plan is complete and the two activities could finish substantially together.
(4) Key-punching revised data could start as soon as the preparation of the revised data started, the two activities finishing simultaneously.

Sketch the network for this situation and analyze it.
(Helpful hint—draw the 'end-to-end' situation *first* and from this draw a modified network.)

Question 20: an exercise in multiple dependencies

Analyze the network shown in Fig. Q20 and draw the appropriate bar chart.

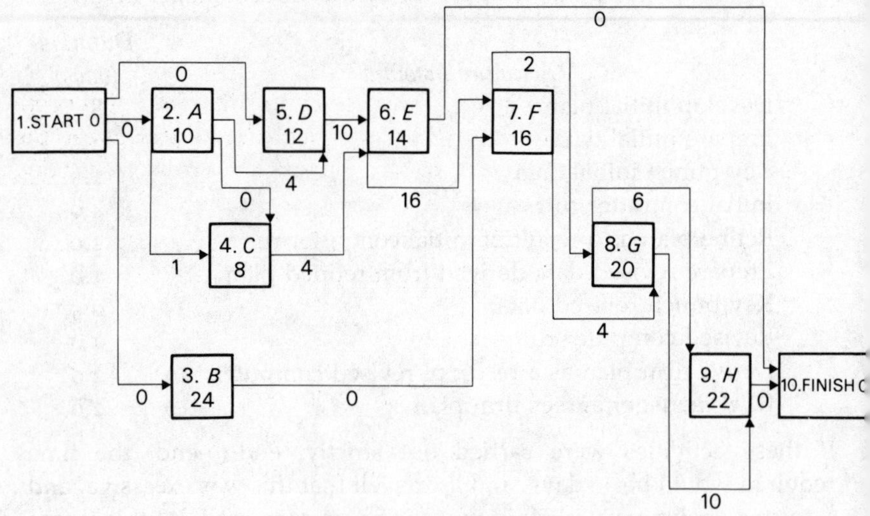

Fig. Q20

Selected reading

BS 4335: 1972: Glossary of terms used in Project Network Techniques (PNT).

BS 6046: Use of network techniques in project management.

Part 1*: Guide to the use of management planning review and reporting procedures.

Part 2: 1981: Guide to the use of graphical and estimating techniques.

Part 3: 1981: Guide to the use of computers.

Part 4: 1981: Guide to resource analysis and cost control.

(* to be published shortly)

These five rather expensive booklets are available from:

British Standards Institution
2 Park Street
London, W1A 2BS

At the time of writing there is an unfortunate difference between recommendations concerning the AoN symbol to be used between BS 4335 and BS 6046: Part 2. The author has found BS 4335 to be of very much greater value than the whole of BS 6046.

Battersby, A. *Network Analysis for Planning and Scheduling* (3rd edn.)
Macmillan, London, 1978.
As with all Albert Battersby's books, this is lucid and readable. While his tragic death has prevented this from being re-written it contains a great deal of valuable information. It includes a chapter on MoP.

Burman, P. J *Precedence Networks for Project Planning and Control*
McGraw-Hill, London, 1972.
This is a highly detailed work of some 370 pages. While slightly dated—computer practice is changed by the use of real-time and interactive computing—it has a wealth of practical experi-

ence derived from the author's experience as a management consultant.

Cleland, D. I. and King, W. R. — *Systems Analysis and Project Management* (3rd edn.) McGraw-Hill, New York, 1983.
While not dealing with the techniques of PNT, this book is useful in that it tries to discuss the complex problems of project management and matrix organization.

Elmaghraby, S. E. — *Activity Networks*
Wiley, New York, 1978.
This provides a thorough discussion of PNT. Anybody wishing to use the 'three-time estimate' method of deriving a duration time is advised to read chapter 3 of this book first.

Fondahl, J. W. — *A Non-computer Approach to Critical Path Method for the Constructional Industry*
The Department of Civil Engineering, Stanford University, California, 1961.
As the title suggests, this work deals with the manual manipulation of networks. It is clear and concise, and its usefulness is far greater than the limitations suggested by the words 'for the constructional industry'. It contains the very first discussion of an activity-on-node (AoN) system.

Harris, R. B. — *Precedence and Arrow Networking Techniques for Construction*
Wiley, New York, 1978.
This is a long book of 420 pages. Very detailed, it uses some symbols that are not in general use—for example, circles for AoN nodes. This is not easy reading but will repay persistence. It perpetuates the 'three-time' myth.

Harrison, F. L. — *Advanced Project Management*
Gower Press, Aldershot, 1981.
This book deals with the managerial problems of project management and spends very little time on the technique of network analysis. Useful for the practitioner.

Lock, D. — *Project Management* (2nd edn.)
Gower Press, Aldershot, 1977.
This is an attempt to consider all aspects of

project management including CPA. Again, many illustrations are from the construction industry.

Martino, R. L. *Project Management and Control vol. 1: Finding the Critical Path, vol. 2: Applied Operational Planning, vol. 3: Allocating and Scheduling Resources*
American Management Association, New York, 1970.
These are three excellent books written by one of the pioneers of this subject.

Moder, J. J. and Phillips, C. R. *Project Management with CPA and PERT* (2nd edn.)
Van Nostrand Reinhold, New York, and Chapman & Hall, London, 1971.
This is a useful book that covers not only the introductory material but also some of the newer methods of allocating resources. The availability and use of some 50 different computer programs is discussed.

O'Brien, James J. *CPM in Construction Management* (2nd edn.)
McGraw-Hill, New York, 1971.
As indicated by the title, all the illustrations and examples are derived from the construction industry. The approach is extremely practical and demonstrates the author's wide experience.

Phillips, D. T. and Garcia-Diaz, A. *Fundamentals of Network Analysis*
Prentice-Hall, Englewood Cliffs, New Jersey, 1981.
The term 'network analysis' refers to a powerful general method of solving many quantitative problems, including project network projects. Thus, some texts including the words 'network analysis' in their titles contain only short and often rather abstract discussions of PNT. One such text is the present work of 470 pages. It contains a very thorough discussion of network analysis with examples of its use in solving, say, the shortest route problem, various LPs, scheduling and so on. One chapter deals with project network techniques quite thoroughly, although some of the terminology is a little unusual.

Schaffer, L. R., Ritter, J. B. and Meyer, W. L. — *The Critical Path Method*
McGraw-Hill, New York, 1965.
This describes both conventional arrow diagrams and the circle diagram technique whereby the use of dummies can be avoided. The use of computers is discussed, and there is a substantial treatment of the 'cost-slope' technique.

Staffurth, C. (ed.) — *Project Cost Control Using Networks* (2nd edn.)
Heinemann, London, 1980.
Principally concerned with cost control, this book includes an outline of PNT, cost control and inflation. Considerable attention is paid to the management problems and implications of cost control. It is well worth studying.

Woodgate, H. S. — *Planning by Network* (3rd edn.)
Business Books, London, 1977.
This is a comprehensive work that derives greatly from its author's experiences within one of the largest of the computer companies. It discusses AoA, AoN and multi-dependency AoN as well as resource allocation and cost control.

Glossary of terms

Wherever appropriate, the term and its definition as given in BS 4335: 1972 are used and appear between inverted commas.

Activity: an operation or process consuming time and possibly other resources.
Activity span: the time available for the completion of an activity.
Activity time: see 'Duration'.
Arrow: the symbol by which an activity is represented.
Arrow diagram: the statement of the complete task by means of arrows.
Backward pass: the procedure whereby the latest event times or the finish and start times for the activities of a network are determined.
Circle: the symbol by which an event is represented in activity-on-arrow.
Cost-slope: the cost incurred in reducing the activity time by unit time. (*Note*: a negative cost-slope indicates that the cost of completing an activity decreases as the activity time decreases.)
Critical path: that sequence of activities which determines the total time for the task.
Critical path analysis (*CPA*): one name for that PNT system wherein an activity is represented by an arrow. Other names which may be used are critical path method (CPM) and program evaluation and review technique (PERT).
Dangle (dangling activity): an activity whose completion does not give rise either to another activity or to the completion of the whole project.
Dependency rule: the basic rule governing the drawing of a network. It requires that an activity which depends on another activity is shown to emerge from the head event of the activity upon which it depends, and that only dependent activities are drawn in this way.
Dummy: a logical link, a constraint which represents no specific operation. In calculations it is most usefully regarded as an activity that absorbs neither resources nor time. (*Note*: dummies are usually represented by broken arrows.)
Dummy activity: a dummy.

'Duration: the estimated or actual time required to complete an activity.'

Duration time: see 'Duration'.

Earliest event time (*EET*): the earliest time by which an event can be achieved without affecting either the total project time or the logic of the network.

Earliest finish time of an activity (*EFT*): the earliest possible time at which an activity can finish without affecting the total project time or the logic of the network.

Earliest independent finish time (*EIFT*): the earliest time an activity can finish without changing the total project time or changing the float in a previous activity.

Earliest start time of an activity (*EST*): the earliest possible time at which an activity can start without affecting either the total project time or the logic of the network.

'Event: a state in the progress of a project after the completion of all preceding activities but before the start of any succeeding activity.'

Event time: the time by which an event can (or is to be) achieved.

'Float: a time available for an activity or path in addition to its duration (may be negative).' It is essentially a property of activities, and is the difference between the time necessary and the time available for an activity.

Forward pass: the procedure whereby the earliest activity times for a network are determined.

Free float: the float possessed by an activity which, if used, will not change the float in later activities.

Free float—early: another name for free float.

Free float—late: the float possessed by an activity when its predecessors and successors are achieved as late as possible.

Head event: the event at the finish of an activity.

Head slack: the slack possessed by an event at the head of an activity.

i: the symbol for the event number of a tail event.

'Imposed date: a date (or point in time) determined by authority or circumstances outside the network, or the fixed point for the time-scale of the network.'

Independent float: the float possessed by an activity which, if used, will not change the float in any other activities in the arrow diagram.

Interface: an activity or event that occurs identically in two or more networks or sub-networks.

Interface (to): the act of coalescing two or more networks.

Interference float: a component of float equal to the head slack of an activity.

j: the symbol for the event number of a head event.

Junction: another name for an event.

Latest event time (*LET*): the latest time by which an event can be achieved without affecting either the total project time or the logic of the network.

Latest free finishing time (*LFFT*): the latest time by which an activity can finish without changing the total project time or changing the float in any later activity.

Latest finish time of an activity (*LFT*): the latest possible time by which an activity can finish without affecting either the total project time or the logic of the network.

Latest start time of an activity (*LST*): the latest possible time by which an activity can start without affecting either the total project time or the logic of the network.

Loop: a sequence of activities in which a later activity is shown to determine an earlier activity.

Method of potentials (*MoP*): a networking system wherein an activity is represented by a node, and dependency is shown by an arrow. Only one type of dependency is used, the start of a succeeding activity being defined by the start(s) of its predecessor(s). The necessary interval between starts is shown as a subscript to the dependency arrow.

Milestone: another name for an event. Sometimes reserved for a major or important event.

Negative float: the time by which the duration of an activity or chain of activities must be reduced in order to permit a scheduled date to be achieved.

Negative slack: the time by which the difference between the earliest and latest event times for an event must be increased in order to permit a scheduled date to be achieved.

'Network: a diagram representing the activities and events of a project, their sequence and inter-relationships.'

Node: another name for event.

PERT: a name for a network analysis technique formed from the words 'program evaluation and review technique'. Originally requiring the use of three estimates of the duration times, the name is now usually accepted as one of the generic names for network techniques.

Precedence diagramming: networking system wherein an activity is represented by a node and dependency is shown by an arrow. At least three types of dependency can be shown, each having different locations on the associated nodes: namely:

Normal (*N*)—the *start* of an activity depends upon the *finish* of its predecessor(s).

Finish-to-finish (*F*)—the *finish* of an activity depends upon the *finish* of its predecessor(s).

Start-to-start(*S*)—the *start* of an activity depends upon the *start* of its predecessor(s).

Project network techniques (*PNT*): the generic term for that group of techniques whereby a project is represented by a set of nodes joined by a set of arrows. It embraces both the system where the activity is represented by an arrow (activity-on-arrow (AoA) technique) and where the activity is represented by a node (the activity on node (AoN) technique).

Resource: anything other than time that is necessary for carrying out an activity. It may be:

Simple (non-storable)—if not used when available is extinguished.

Pool (storable)—may be used when required.

'Resource aggregation: the summation of the requirements of each resource for each time period, calculated according to a common decision rule'

Resource levelling: the utilization of the available float within a network to ensure that the resources required are appreciably constant.

'Resource limited scheduling: the scheduling of activities such that predetermined resource levels are never exceeded, and the project duration minimized.'

Resource optimization: the manipulation of the network to try to ensure that the resources required and available are in balance.

'Resource smoothing: the scheduling of activities within the limits of their total floats, such that fluctuations in resource requirements are minimized.'

Resource totalling: equivalent to 'resource aggregation'.

Scheduled date
Scheduled time } equivalent to 'imposed date'.

Secondary float: when a scheduled date is imposed upon an activity which is not a final activity, a secondary critical path can appear which is the time-controlling sequence between the start and the scheduled event, or between two scheduled events. Activities not on this critical path but which contribute to the achievement of the event possess float with respect to this secondary critical path, and this is said to be secondary float.

Semi-critical path: that path which is next to the critical path when all paths are arranged in order of float.

'Slack: Latest date of event minus earliest date of event (may be negative). The term slack is used as referring only to an event.'

Stage: another name of an event.

Sub-critical path: a path which is not critical.

Tail event: the event at the beginning of an activity.

Tail slack: the slack possessed by an event at the tail of an activity.

Time available: another name for 'activity span'.

'Time limited scheduling: The scheduling of activities such that the specified project duration is not exceeded, using resources to a predetermined pattern.'

Total float: the total float possessed by an activity.

'Trading off': the transferring of resources from one activity to another. This is usually accompanied by changes in duration times, and is carried out to affect the resource distribution.

Index